Reflections Underwater

Published in 2023 by
Pelagic Publishing
20–22 Wenlock Road
London N1 7GU, UK

www.pelagicpublishing.com

Reflections Underwater
A Multidisciplinary Exploration of Coral Reef Wonders

Editor: Shoshana Brickman
Scientific editor: Dr. Tom Shlesinger
Photographs: Oded Degany
Illustrations: Anath Abensour
Book design: Inbal Reuven

Additional credits are included on the bibliography and figure credits pages.

https://doi.org/10.53061/PJYM7953

British Library Cataloguing in Publication Data
A catalogue record for this book is available from the British Library

ISBN 978-1-78427-413-9 Hardback
ISBN 978-1-78427-414-6 ePub
ISBN 978-1-78427-415-3 PDF

Printed in Czech Republic by FINIDR

www.oded-degany.com | reflections.underwater2021@gmail.com

Front cover: Pinnacle of Crinoids (sea lilies). Puerto Galera, Philippines.
Spine: School of anthias. Verde Island, Philippines.
Back cover: Spotted-ribbontail ray's eye. Eilat, Israel; Freckled (scarlet) frogfish. Lembeh Strait, Indonesia.

Reflections Underwater

A Multidisciplinary Exploration of Coral Reef Wonders

Oded Degany

Foreword by Professor David Fortus

Afterword by Dr. Tom Shlesinger

Pelagic Publishing

This book is dedicated to my father,
Jacob Degany (1932–1994),
who could not see it come to fruition.

"The sea is everything. It covers seven tenths of the terrestrial globe. Its breath is pure and healthy. It is an immense desert, where man is never lonely, for he feels life stirring on all sides. The sea is only the embodiment of a supernatural and wonderful existence. It is nothing but love and emotion; it is the Living Infinite."

Jules Verne, *Twenty Thousand Leagues Under the Sea: A World Tour Underwater*, 1872

« Sea turtle. Bunaken, Indonesia.

Contents

"Shuzan held out his short staff and said: 'If you call this a short staff, you oppose its reality. If you do not call it a short staff, you ignore the fact. Now what do you wish to call this?'"

Zen Koan

Foreword

By Professor David Fortus

The Chief Justice Bora Laskin Professorial Chair
of Science Teaching, Weizmann Institute of Science

« Foraging lionfish. Eilat, Israel.

There is something on my table. It is almost symmetrical around a vertical axis. It is hard, but not brittle. It is made of organic materials. Once it was home to many organisms, now to fewer. It is wet on the inside but dry outside. It does not conduct electricity. It has sentimental value to me. What is it? Do these descriptions of the object's properties bring it to life or do they actually deaden it? Can knowledge of these properties replace the sense of appreciation that may come with experiencing the object, by holding it, seeing it, using it, smelling and perhaps tasting it? Do they limit or enhance one's experience with the object?

Whenever we inspect any object or phenomenon, we see different things depending on the perspective we adopt. A physicist sees in a leaf countless energy transformations, a chemist the synthesis of sugars, the evolutionary biologist an adaptation to an environment, an indigenous shaman the source of a substance to relieve pain, and so on... So is a leaf just a leaf, or is it much more? Can we really know what a leaf is and fully grasp its reality, or by simply calling it a leaf are we actually constraining it and limiting our ability to fully appreciate it?

Epistemology, which deals with the nature and the justification for our perceptions and conceptions of the world, is central to philosophical discourse. Emmanuel Kant was the first to challenge the longstanding historical pretension that the world could be understood in a completely objective manner, claiming that our conceptions of the external world were not accurate reflections of objective external reality. Instead, according to Kant, the human mind acts as an organizer, constructing our conception of reality based on a priori concepts and our senses' inputs and not merely on experience. A priori concepts and sensation, Kant believed, led to conception, and different senses would lead to different conceptions. Since our senses are uniquely human, we have a unique and anthropocentric conception of nature. We simply cannot view the world other than the way we do. This anthropocentrism may underlie our ethical approach to nature, leading to the belief that humans stand in the center of the universe and are superior to nature, a worldview that is criticized several times in Degany's book. However, Kant's theory has itself been challenged; the fact that Degany is human but argues against anthropocentrism can be seen as evidence that not all our conceptions may be derived from our senses.

This remarkable book clearly adopts the perspective that knowledge enhances experience, but cannot replace it. Reading this will not make diving at coral reefs unnecessary; it will embellish the experience. If you dive regularly at coral reefs, this book will show you new ways of seeing the reef, changing your experience of it. But rather than having to read one book about reefs by a physicist, another one by a mathematician, another one by a marine biologist, another one by an anthropologist, and yet another one by a philosopher, Degany has done something unique, he has combined all these perspectives in a single book. He has succeeded in demonstrating that, as far as the reef is concerned, everything is connected.

But more than that, Degany continually draws parallels between the reef and the human condition, so you will learn a lot about people, not just about polyps and octopuses and other marine creatures. You will learn about reproduction strategies in the reef and have these compared to gender and sexuality in humans. You will learn about the power of numbers, not only in fish schools, but with people. You will learn about shamanism, zombies, the dispute between Newton and Goethe about color theory, and much more. In the end, this book provides a very humanistic perspective on the reef: by learning about the reef, we are learning about ourselves.

In the end though, no matter how much we learn, we are limited in our ability to conceive of the reef in its totality, just as I am limited in my ability to understand all the various aspects that make the object on my desk what it is. In spite of the incredible eruditeness underlying this book, it is clear to Degany that he is only scratching the surface, that some things cannot be fully explained by reductionist approaches like science. In this sense, this book is modest, about our ability to understand nature and about humanity's place in nature. Hopefully, this will be another insight that readers will take away with them: We are blind elephants in an intricately organized china store. Even when our eyes are opened, we can still barely see. Let us be aware of our limitedness and proceed cautiously, showing respect for and awe of Mother Nature.

This book is extraordinary, original, beautiful, and fascinating. Degany has an incredible mind. I have met many brilliant people, but only extremely few have interests so broad, are knowledgeable about such a wide range of disciplines, and delve into the tiniest details underlying a topic but at the same time are able to see the big picture. This book justifies Descartes, who thought that "The reading of all good books is like a conversation with the finest minds...." At a time when society encourages, promotes and often dictates specialization at the expense of broadness, when too many scientists know everything about almost nothing but almost nothing about everything, it is refreshing to see that it is still possible to be both broad and deep. It is also refreshing to see that there are fundamental questions that cannot be addressed without using multidisciplinary perspectives.

While this book is about coral reefs, it is really about much more than that. Coral reefs are only a theme to hold everything together. In reality, this book is about physical and geometric optics, evolution, zombies, mathematics, and the occult. It is also about philosophy, Turing's pattern theory, symbolism, and superorganisms. Add to that incredibly beautiful photographs and you have an excellent book that will continually surprise you.

This book is not always easy to read. At times, it may seem overwhelming. But it will definitely make you think. It will make you consider possibilities that you might have otherwise rejected. Most of the ideas presented in the book are firmly grounded in science, but some of them are controversial. This book floats somewhere between the known and the possible. It does its best to take a step out of the box, while keeping one leg firmly planted inside the box. Degany is apparently a "what if" person. He accepts nothing as an absolute truth. To paraphrase Voltaire's saying in this context, he recognizes that, while uncertainty is unpleasant, certainty is usually ridiculous.

This is a book that you should read if you want to be amazed anew at the richness of the coral reef, to discover new ways to think of it, to see it as you probably have never seen it before. You don't have to read this book from start to end without a break. You can read each chapter as if it were independent of the others. The various chapters do not have to be read in any particular order. Look at the table of contents, select a topic that intrigues you, and start there. I guarantee that you will be fascinated and learn many things.

Good educators know that one of their goals is to expose their students to unfamiliar phenomena and stories that they would be unlikely to encounter on their own, to provide them with experiences that will fire their imagination, to confront them with new ideas, new ways of thinking. Until I read this book, I never thought of coral reefs in such a broad context and in such an interdisciplinary way. Sure, I knew that coral reefs are beautiful and rich ecosystems, but this book's ability to describe the science underlying the reef in a humanistic way is refreshing and intriguing. This book changed my perception of reefs. I am sure it will do the same for you.

Israel
March 2021

Acknowledgments

Writing a book is more challenging than I initially thought, and more rewarding than I could have envisaged. From the first concept in my head to the complete manuscript, I have passed through a lengthy mental and intellectual process. Throughout this journey, many people were involved; all of them deserve to be acknowledged and thanked for helping me bring this book to fruition.

This book results from endless discussions I have conducted with numerous people about ideas that intrigued me over the years. While they all revolve around coral reefs, these ideas broadly reflect my inner world and eclectic character.

First and foremost, I would like to thank numerous scholars who have dedicated their lives to studying the topics covered in this book. Each and every one of these scholars stands on the shoulders of giant thinkers who are responsible for the intellectual progress throughout the history of human ideas. Their names, and my debts to them, fill this book.

I owe an enormous debt of gratitude to those who inspired me and offered me invaluable help on one or more chapters or photos, including Eli Amit, Peninah Brickman, Guy Degany, Natan Dotan, Hagit Arad, Prof. Maoz Fine, Prof. David Fortus, Sami Gaist, Dimpy Jacobs, Dr. Amit Lotan, Michal Peer, Dror Schnitkes, Dr. Sara Schwartz, Hanan Shmuely, Prof. (Emer.) Yehuda Shoenfeld M.D., Carmel Vernia, and Ram Zeevi. I also want to thank all of those who buoyed my spirit and were part of my journey and are not mentioned explicitly here.

I want to thank Dr. Ramy Klein, who was part of my diving journey, for his intellectual generosity, sharing his immense marine biology knowledge, and introducing me to some of the coral reef's most mysterious creatures.

Additionally, my special thanks to Dr. Tom Shlesinger, a promising young marine biologist who has already attained significant scientific achievements. I thank Tom for his scientific editing and also for his insightful advice and comments regarding humanistic topics covered in this book. Tom belongs to a rare category of scientists who are not just the highest-level expert in their field but also interested in broad scientific and humanistic ideas.

Enormous appreciation I owe to my editor Shoshana Brickman, who closely accompanied me along the journey and was part of endless intellectual discussions around the book's topics. Rational thinking, professionality, wisdom, diligence and endless curiosity are only a few of her merits. Obviously, she was the best colleague

I could have asked for. Shoshana's professional support and intellectual contribution to the writing process were immense and are engraved throughout the book.

I also want to thank my book designer, Inbal Reuven, and my illustrator, Anath Abensour, for their creativity and lovely designs.

I am grateful to Nigel Massen, David Hawkins, Sarah Stott, and everyone at Pelagic Publishing, and to Chris Reed and Amanda Thompson at BBR Design.

Needless to say, all errors and mistakes in the book remain my own.

My closest partner in the writing of this book was my son, Or, who is also my preferred diving partner. Or's endless curiosity, scientific knowledge, and creative ideas inspired me and are thoroughly integrated into the book.

I want to thank my ever-patient wife, Zohar, and my daughters, Eden and Gal, for tolerating my disappearances into my home office. I also want to thank my beloved dog, Whiskey, for his companionship and patience during the writing process. His sphinxlike gaze induced calmness, wisdom, and thoughtfulness. Lifelong partners are a prerequisite for success and what makes the journey so enjoyable and meaningful.

Finally, I dedicate this book to my mother, Lilia, and my late father, Jacob, who instilled in my heart the love of human culture, and who encouraged me to adopt an open-minded and inquisitive approach.

Oded Degany
November 2022
www.oded-degany.com | reflections.underwater2021@gmail.com

Prologue

By Oded Degany

« Green sea turtle. Bunaken, Indonesia.

My journey to the underwater wonderland of coral reefs began when I was a child, during my first visit to the coral reefs of the Red Sea. It was the late 1970s, the area was under Israeli rule, and the beautiful shores of the Sinai Peninsula were a popular destination for Israeli tourists. Sinai was a place at the margins, extremely remote and relatively isolated from the rest of Israel – a heterotopia, to use a term coined by Michel Foucault (1926–1984). I remember the first time I immersed myself in the crystal-clear blue water of a reef called Ras-Burqa. I was mesmerized by the magnificent colors illuminated by the intense desert sun. I had grown up in a Mediterranean coastal city and had never encountered a coral reef before, so I was stunned by the new scenery.

This unforgettable childhood impression emerged many years later, during the summer of 1999. At the time, I was exhausted from my work as an M&A executive in one of the largest holding companies in Israel and decided to replenish my energy by taking a scuba diving course. In my imagination, deep-sea diving would provide me with a temporary refuge or illusory freedom from the mundane world.

It was an open-water diving course in Tel Aviv, on the shores of the East Mediterranean, where water visibility is poor, the sea is choppy and, from a layperson's perspective, underwater marine life is minimal. Needless to say, the experience was not a memorable one.

My decision to pursue an advanced open-water course came mainly from the memory of my childhood experience in the Sinai Peninsula. Deep within me, I remembered the view of the reef, the calmness induced by seeing the reddish mountains meeting the deep blue sea.

Since then, I have expanded my diving destinations and spent my holidays diving in different regions worldwide, mainly in the Asian Pacific, engaging with the sensual and intellectual experience of diving in tropical coral reefs.

My academic background in physics and the philosophy of biology provided me with an intellectual layer of enjoyment, enriching my entire experience. After several years of diving, I added another layer and began integrating photography into diving.

An additional factor that drove me to further explore the wonders of the underwater world was an accidental encounter I had with a film by Leni Riefenstahl (1925–2002). Riefenstahl, a German actress and film director notorious for her role in Nazi propaganda, was a pioneer in documentary propaganda, having directed films such as *Triumph of the Will*, about the Nazi Party Congress in Nuremberg, and *Olympia*, about the 1936 Summer Olympics, both of which glorified Nazi party values and aesthetic ideals. In her final film more than 50 years later, *Impressionen unter Wasser*, Rienfenstahl focused on the remarkable aesthetics of coral reefs and the splendor of nature. Riefenstahl's film was unique, since unlike other nature documentary films of their era, it was pure artwork. Famous for her photography techniques, she filled the film with graceful images of coral reef scenery and vibrant displays of marine creatures. I remember the opening of the film, for example, where she invites the viewer to enter the underwater kingdom through the coiled structure of a wire coral.

Diving is a broad hobby, one which people are drawn to for diverse experiences. Some divers are interested in the pelagic world, observing sharks, whales and large shoals of fishes. Dive sites such as Cocos Island and the Galapagos enable fans of pelagic animals to encounter big schools of those animals. Personally, I prefer the colorful and vivid nature of coral reefs. There is no contradiction between the two, but eventually, if diving becomes a significant hobby, divers tend to define what they like the most.

Over the years, I have dived in many places, including the Red Sea, the Andaman Sea, the Coral Triangle (an area surrounding Indonesia, Papua New Guinea, and the Philippines), the Caribbean Sea, and the Pacific Ocean. Over time, it has become clear to me that dive sites in the Philippines and Indonesia are more abundant and more colorful than other reefs around the world. Moreover, in addition to the large number of species, from all different animal groups, it has also become clear that the number of encounters with each species is higher here than in other spots. Places like Puerto Galera and Dumaguete in the Philippines and Lembeh Strait, and Bunaken Marine Reserve in Indonesia, provide the best diving experiences on the planet, with an abundance of marine life and diving opportunities.

Some divers and underwater photographers, including me, favor muck diving, a genre of diving named for the muck or sediment at the bottom of the dive sites. Many muck diving sites are located in tropical areas and often include human-made garbage, wrecks, and piles of broken dead coral. These dive sites can be extremely rich in marine animals such as shrimps, crabs, sea slugs, octopuses, and many species of fishes. Muck divers do not enjoy a colorful coral reef, but the cornucopia of marine life compensates for the lack of coral reef beauty.

This book is inspired by my eclectic, multidisciplinary adventure into coral reefs. Throughout my diving journey, my sensual experience is accompanied by thoughts about the guiding principles of evolution theory and an attempt to connect life in the coral reef to universal scientific and humanistic foundations. The remarkable variety of living forms, and the way they manifest biological principles in coral reefs, shattered the traditional limits of my mundane thoughts and provide me with an ongoing, limitless adventure. I am a curious person, one who cannot avoid experiencing the world intellectually, trying to find the fundamental mechanism behind every phenomenon of life. Physicists call this the quest for the Theory of Everything. For me, it is clear that in the science of systems such as biology, it is impossible to find answers within a single grand unified theory; however, the aspiration to find basic universal rules is, in itself, what makes the intellectual journey so fascinating.

We are living in a world that sanctifies specialization and narrow thought (in the best case). This book advocates for the opposite: that a multidisciplinary approach to nature's phenomena provides an enriched angle that cannot be obtained by a single-discipline scientific point of view. In my opinion, a multidisciplinary approach reflects the nature of life, which is an eclectic mixture of many sorts of things that can be viewed from multiple angles and by using unrelated perspectives. Personally, I follow British biologist and anthropologist Thomas Huxley (known as "Darwin's Bulldog"), who coined the expression, "Try to learn something about everything and everything about something." If your senses, mind and heart are open, coral reefs can serve as an enlightening prism to many phenomena of life.

Moreover, people tend to think that human perception is a passive objective process. Human perception and knowledge form an active constructive process based on human senses and, no less importantly, cultural context. Our perceptions are skewed by preconceptions, experiences, and expectations. Inspired by this notion, I decided to weave humanistic themes that relate to underwater marine phenomena into this book.

Coral reefs provide visitors with a unique concentration of mysterious phenomena. They occupy a tiny portion of the ocean's seabed but host an incredible and disproportionate amount of its biodiversity. Moreover, for millions of years, they have served as an evolutionary cradle and source of biodiversity for non-tropical areas, too.

Huayan, meaning flower garland, is a Buddhist school of thought that thrived in China during the Tang Empire (618–907 CE). The central concept in *Huayan* Buddhism is ultimate unity, reflected in the phrase "one is all, and all is one." The recognition that we are all parts of one transpersonal reality eventually leads to the concept of universal compassion.

One of the metaphors from the *Avatamsaka* Sutra, one of the most influential *Mahāyāna* sutras in the *Huayan* School, is Indra's net, a metaphorical net that has a jewel at the intersection of every two cords. Each jewel is so bright that it reflects the qualities of every other jewel in the net, similar to a hologram, in which every point contains information about all other points. Looking into one, you see them all. When any of the jewels in the net are touched, all the other jewels are affected. For me, Indra's net embodies the eternal interconnectedness and interdependency of the animals of the coral reef.[1]

We live in a time of unprecedented urgency with respect to global climate change. I hope that *Reflections Underwater* inspires people to dedicate more resources and attention to global threats and their consequences on coral reefs. If we do not care and act now, future generations might not be able to experience the beauty of the world as we do today. I deeply hope that sharing my awe of the reef's mysteries and splendor will be my humble contribution to the conservation of precious coral reef habitats.

How to read this book

Not a zoological textbook, *Reflections Underwater* explores meta-themes beyond the general systematics of biology, combining the disciplines of philosophy, psychology, evolution, art, zoology, physics, mathematics and more. My purpose is to highlight coral reefs as a unique region of wonder, where awe is experienced at every moment. I have tried to avoid going into taxonomical details unless they relate to a thorough description of a phenomenon or a particularly intriguing anecdote.[2] When writing about generalizations, I may avoid mentioning the exceptions. Biology and ecology are sciences of systems, and any attempt to describe them in a reductionist manner is doomed to fail.

This book is for people who are curious, or nature enthusiasts, or divers or snorkelers, who want to stimulate their thinking by considering multidisciplinary topics that relate to coral reefs. I have taken the liberty of discussing only topics that I think are of interest, without providing a systematic overview of each subject. The scarlet thread of the book is the coral reef and its association with scientific and

1 https://plato.stanford.edu/entries/buddhism-huayan/ (retrieved March 6, 2021).

2 Taxonomy in biology is the systematic classification of organisms.

humanistic ideas. The book is also a broad way of looking at the phenomena of life through the unique prism of coral reefs.

Due to the abundance of ideas and data in this book, and in order to avoid information overload, you are invited to browse through the book and discover it as you want. There is no need to read it from the first page to the last. It is structured in a self-explanatory manner that enables you to read most of the chapters and sections thereof independently. I have prepared a short, friendly list of basic terms to assist anyone who is not familiar with the common terminology of the topics covered in the book. To those of you who are new to the basic terms of coral reef biology, I suggest reading this short glossary first.

Photographs that I have taken throughout my years of diving are integrated into the book and accompany the major themes discussed. I hope that they manage to convey the remarkable beauty and wonder that lies within coral reef ecosystems.

The structure

Chapter 1 focuses on the psychology behind the sensual and intellectual diving experiences. I explore several of the major theories, most of them oriented towards cognitive evolutionary psychology, that suggest reasons for our attraction to water and the uniquely sensual experience of diving. In this chapter, I advocate for the complementary value of the intellectual experience in diving. For me, the intellectual and sensual experiences are intertwined, providing a complete experience of reef life. I also address the postmodern and cognitive psychological views of photography, especially in the context of my underwater photography experience.

Chapter 2 discusses the coral reef as an entire ecosystem. The chapter starts with a basic description of the characteristics of coral as an animal, including coral reproduction and its fundamental symbiosis with algae. Later, I discuss topics that share a common denominator, which is a systems view of coral reefs. These topics include the reasons for their astonishing biodiversity, our biased perspectives of the healthy reef myth, questions regarding the reef's behavior as a superorganism, a possible solution to Darwin's Paradox, and more.

Chapter 3 addresses one of the major riddles regarding coral reefs: Why are they so colorful? This chapter deals with the high-level constraints set by natural selection and how reef animals find the "golden path" of addressing contradictory demands. The chapter also covers the properties of light in the marine environment and how those properties influence reef animals' coloration. Camouflage is also an essential factor in animal coloration. Some of its principles are non-intuitive and used by artists as well as the military. The last point of this chapter deals with human bias towards specific colors and other factors that dictate how we view the reef.

Chapter 4 describes the exciting vision and color creation mechanism of coral reef animals. Fluorescence and bioluminescence are two of the ways reef animals produce light and color. I also describe the blue-ringed octopus and disco clam's unique ways of producing warning color and light. The vision systems of reef animals are diverse and include a wide array of methods for sensing light and "seeing" the environment. One of the most astonishing mechanisms is the eye of the mantis shrimp, considered to be the most sophisticated eye in nature. I challenge this opinion, presenting its advantages and disadvantages compared to other types of vision systems. The last part of this chapter deals with the mind-boggling disguise mechanism of the octopus and a possible solution to the curious puzzle regarding its colorblindness.

Chapter 5 covers one of the most fundamental phenomena of the reef, the symbiotic relationships between various reef dwellers. A coral reef is an interconnected system. As such, it exemplifies, along with all the forms of life within it, the universal concept of holism[3] and mutual dependency between its parts. Symbiosis and collaboration between animals are common themes in this book and represent my belief in the importance of cooperation as a driving force in evolution and a source of evolutionary novelty. Alongside the remarkable examples of symbiosis in the reef, I also address, in the case of cleaning stations, the non-trivial evolutionary question of how those symbiotic relationships have evolved.

Chapter 6 explores unique preying and defense strategies and mechanisms found in the reef. One of these belongs to the category of ultrafast biological systems that demonstrate performance far superior to that of human-made machines. This includes the deadly strikes of the mantis shrimp and the hunting cells of the cnidarians (animals that possess stinging cells, including corals, sea anemones, jellyfishes and more). Alongside cryptic behavior, venom is a common defense mechanism in the reef. This chapter describes various venomous reef creatures, the

3 Holism is the idea that various systems (human cultures, physical, biological) are more than just the sum of their parts. In other words, understanding the behavior of the individual parts is insufficient for understanding the entire system behavior.

coevolution of venom and warning coloration, the self-organization property of fish schooling, and its merits as a defense mechanism.

Chapter 7 deals with reproduction in the coral reef, which is entirely different from what we are familiar with as terrestrial mammals. In this chapter, I discuss topics such as sex change in fishes, hermaphroditism, the puzzle of hybrid animals in light of the traditional definition of species, the unique fatal mating act of octopuses, nudibranch mating, and more.

Chapter 8 describes the new paradigm of evolution theory, which significantly diverts from what Darwin and Wallace thought in the mid-nineteenth century. The public discourse about evolution includes many misconceptions, and causes people to adopt hasty conclusions about the phenomena of life. I also review topics such as animal consciousness, in general, and octopus consciousness, in particular, aiming to present a different perspective on human–nature relationships. I strongly believe that changing our traditional perspective about animal consciousness can have a positive effect on how we perceive nature. This chapter also covers an intriguing topic which diverts from the prevailing paradigm regarding the source of life on Earth and the hypothesis regarding the possible extraterrestrial origin of octopuses. In addition, I cover the puzzle regarding the evolutionary source of eyes, and the enigma of warning coloration evolvement.

Chapter 9 provides a glimpse of the mother of all sciences: mathematics. In this chapter, I discuss the manifestation of the universal laws of mathematics in coral reefs, including the golden section, symmetry, hyperbolic geometry, and the coral reef as a complex dynamic system.

"The Buddha, the Godhead, resides quite as comfortably in the circuits of a digital computer or the gears of a cycle transmission as he does at the top of a mountain or in the petals of a flower."

Robert Pirsig, *Zen and the Art of Motorcycle Maintenance: An Inquiry into Values*, 1974

CHAPTER 1

The Uniqueness of the Diving Experience

Water Symbolism

« A school of blotcheye soldierfish in an entangled coral reef. Eilat, Israel.

Water is one of the key symbols of human culture and is one of the most recurring motifs in legends, magic, and religions. Considered the source of all potentialities of existence, it is equated with the term *panta rhei*, which means that all things are in flux.[4] Water symbolizes unconsciousness and forgetfulness; it is a means of purifying, dissolving and regenerating, a feminine symbol associated with fertility, refreshment and the fountain of life.

Since water is essential to most forms of life, it is not surprising that water is sanctified by most cultures. As described by historian and philosopher of religion Mircea Eliade (1921–1986) in his 1959 book *The Sacred and The Profane: The Nature of Religion*:

> *The waters symbolize the universal sum of virtualities; they are* ***fons*** et ***origo****, "spring and origin," the reservoir of all the possibilities of existence; they precede every form and support every creation.*

Throughout history, humans have assigned healing, purification and life-changing properties to water. Baths and immersion are central in many cultures: *punyahavachanam* in Hinduism, *tvilla* in Judaism, baptism in Christianity, *onsen* hot spring baths in Japan, *thermae* in ancient Rome, and *hammam* in the Ottoman Empire.

In Ethiopia, for example, where water is highly sacred, it is commonly believed that holy water can cure all types of diseases. Even today, many Ethiopian people

4 *Panta rhei* is an expression attributed to the Greek philosopher Heraclitus.

believe that the healing powers of water are superior to those of conventional Western medicine.

^ An Ethiopian Orthodox Christian pilgrim is baptized in the Jordan River, near Jericho. Qasr-el-Yahud, Israel.

Water symbolism is also reflected in the theory of archetypes put forth by psychoanalyst Carl Gustav Jung (1875–1961). Jung presented the concept of archetypes, the analog of physical instincts, as impersonal, innate archaic images, or universal thought-forms that influence the feelings and actions of an individual, known as the collective unconscious. Archetypes unconsciously motivate human feelings, dreams and behavior throughout our entire life. For Jung, the archetype of water mainly symbolizes the unconscious. From this perspective, the rich underwater world, barely visible from the outside, is revealed in every dive and can serve as a metaphor for the Jungian unconscious archetype. Psychoanalytically, this might shed light on one of the fundamental causes behind the human attraction to diving into the unknown.

Compared with the symbolism of water, the symbolism of the sea is much more ambiguous. On the one hand, the sea is awe inspiring, considered to be the *anima mundi* (the world soul); on the other hand, it is regarded as an unknown *mare incognitum*, dangerous, chaotic and inexhaustible. In *Moby Dick*, Herman Melville (1819–1891) described the dualistic nature of the sea and land as follows:

> *consider them both, the sea and the land; and do you not find a strange analogy to something in yourself? For as this appalling ocean surrounds the verdant land, so in the soul of man there lies one insular Tahiti, full of peace and joy, but encompassed by all the horrors of the half-known life. God keep thee! Push not off from that isle, thou canst never return!*

Tropical coral reefs, which lie at the heart of this book, are entirely associated with the positive aspects of the sea. They are a contemporary symbol of beauty, richness, vividness, and calmness.

‹ Ascidians (sea squirts) are marine invertebrates with early primitive signs for vertebrate features that filter food from the water through two large pores. One pore guides water into the body cavity while the other provides the exit. Dauin, Philippines.

Sensual Experience and the Enigma Behind Our Attraction to Diving

First and foremost, diving is a unique sensual experience, mostly because it places you in an entirely different environmental realm. Akin to other sensual and subjective experiences, diving is challenging to capture in writing, for the same reason that words always degrade real-life experiences. French psychoanalyst Jacques Lacan (1901–1981) asserted that the lived experience is irreducible to the symbolic order. Based on Lacan's doctrine, language has limited capacity for describing reality. Together with the symbolic order, it creates a wall that prevents us from accessing "the real." Only trauma, awe-inspiring (sublime) experiences, or altered states of mind that are mentally, physically, emotionally, or chemically induced can break the barrier formed by symbolism and language, providing us with a glimpse of the real.

Although it is difficult to describe the diving experience, we can still ask the question, "What is the source of our deep attraction and fascination with water and diving?" This can only be answered by incorporating multidisciplinary perspectives, which I address with an emphasis on evolutionary psychology theories.

Alongside the air that we breathe, water is the most omnipresent substance on the planet. Life is built around a very simple molecule, H_2O, which can be found in almost all habitats. Water covers more than 70 percent of the Earth's surface and accounts for approximately two-thirds of the human body mass.

A human can survive for more than three weeks without consuming food. Indian leader and ethicist Mahatma Gandhi (1869–1948), for instance, went on a hunger strike for 21 days. However, surviving without water is another story; a typical human can endure no more than three to four days without water, which demonstrates how utterly dependent on it we are.

Innate human diving skills

The idea that humans evolved in an aquatic habitat is rooted in the history of ideas, and is not part of today's dominant scientific paradigm. Similar to other semi-aquatic animals, humans are well equipped physiologically for diving 20 to 30 meters underwater. Furthermore, there is evidence that humans have used sea harvesting as a primary source of food since the times of *Homo erectus* around two million years ago.

These facts suggest selective pressure for diving during the course of human evolution. Some scientists proposed a competing (or in some cases complementary) theory to the savanna hypothesis, known as the aquatic ape hypothesis (AAH).[5] The AAH is a group of theories that assert that our ancestors went through an aquatic phase, a stage that might explain the differences between humans and other apes. It is suggested that a few million years ago, several species of primitive apes were forced by competition to hunt and feed on creatures at seashores and water bodies. The aquatic ape hypothesis is based on observations that non-terrestrial mammals such as whales share a suite of morphological and physiological features, commonly regarded as being unique to humans.[6] A surprising number of similarities, such as the loss of body hair, subcutaneous fat, aquatic babies, and more, are not unique to humans and might serve aquatic functionality. Although the scientific community currently rejects the AAH, in recent years some voices have called for its reconsideration.

Humans living in an aquatic environment can be seen today in cultures of harvesting divers, which include the nomadic hunter-gatherer indigenous peoples known as sea nomads, who base their diet on seafood obtained through free-diving techniques. These cultures, some of which are as much as 2,000 years old, include the Ama in Japan, the Hae Nyo in South Korea, the Moken along the coasts of Myanmar and Thailand (the Andaman Sea), and the Orang Laut in Indonesia. Harvesting divers may spend 60 percent of their day underwater and have extraordinary underwater vision and breath-holding abilities that enable them to dive to a depth of over 70 meters. Moken children have superior underwater vision, with underwater acuity more than twice that of European children. Research conducted by Dr. Melissa Ilardo and her colleagues on the Indonesian Sama-Bajau, also known as the Bajau, who have engaged in breath-hold diving for hundreds of years, suggests that genetic selection results in an increase of spleen size, providing them with an extra reservoir of oxygenated red blood cells. This research also finds evidence for a potentially superior adaptive diving reflex.[7]

I remember a BBC documentary showing a Bajau man walking underwater (for several minutes!) while hunting.[8] The diving skills of a sea nomad look completely different compared to cumbersome movements of the average scuba diver and can serve as a model of how to merge and live in harmony with nature.

5 The savanna hypothesis postulates that, due to climate change, humans abandoned an arboreal lifestyle and migrated to the savannas. During this transition period, humans evolved bipedalism locomotion.

6 At a later stage, mammals such as seals, dolphins and whales left the land and returned to their aquatic habitat.

7 The mammalian diving reflex is a set of physical responses to submersion in water. These changes enhance survivability while diving by overruling basic homeostatic reflexes. The diving reflex results in an efficient distribution of oxygen to the heart and brain, slow heart rate, hypertension, bradycardia and the release of red blood cells stored in the spleen.

8 https://youtu.be/JSU9RCaiorQ (retrieved March 6, 2021).

Water and coral reef scenery

Another sensual aspect of the diving experience comes from theories that try to explain human aesthetic perception and our attraction to nature. Philosopher Denis Dutton (1944–2010) opposed the view that aesthetics are culturally learned and instead argued that aesthetic sense is the result of evolutionary adaptation. He suggested the Darwinian theory of beauty, which postulates that humans share preferences when they are asked to describe a beautiful landscape. Dutton defined those preferences as a "universal landscape" that includes all the elements needed for human survival. Water is one of those preferences, as is short green grass, open space with trees, and more. The main concepts of Dutton's theory were not developed in a vacuum; they are based on the foundations of sociobiology, as notably described by the father of sociobiology Edward Osborne Wilson (1929–2021). Wilson asserted that there is a genetic, "hard-wired" basis behind humans' attraction to the natural world and our desire for contact with it.

The effect of water scenery on our minds is part of broader research about the phenomena of awe and wonder. Coral reefs, without a doubt, belong to the category of awe-inspiring landscapes. Professor Dacher Keltner (1962–), director of the Greater Good Science Center in Berkeley, defines the sublime as a "feeling of being in the presence of something vast that transcends your understanding of the world."[9] Awe can be a daily experience that is amplified in the presence of extraordinary natural landscapes such as coral reefs. Keltner explains the evolutionary source of awe in human beings by arguing that awe strengthens our social collectiveness, feelings, and social identity. Awe-inspiring experiences encourage people to collaborate and feel empathy towards others. While encountering the wonders of the world, we calibrate our self-importance and embrace a new perspective about our place in the universe. Our self-perception changes and becomes part of something greater than us; we feel part of a greater whole. Furthermore, momentary sublime experiences instill energy, stimulate wonder and curiosity, and inspire us to think creatively.

As with landscapes, and perhaps even more so, it appears that human contact with animals affects human welfare and promotes physiological health and emotional well-being. For example, it was found that gazing at an aquarium is as effective as hypnosis for relaxation, decreasing blood pressure, and increasing comfort levels during surgery. By providing people with relief from life's usual concerns, nature can play a vital role in positively building and managing a sense of self-identity.

---

9 https://greatergood.berkeley.edu/article/item/why_do_we_feel_awe (retrieved March 6, 2021).

The effects of water immersion

The influence of water on the human mind has been studied by marine biologist Wallace J. Nichols, who explores the neuroscience behind the therapeutic effect of water and its positive influence on our minds. He suggests that when we are surrounded by or immersed in water, our minds enter a mildly meditative mode he calls "blue mind." In Nichols's view, water induces a meditative state which is associated with relief from anxiety and depression, and increased awareness and optimism.

Likewise, research conducted by Frédéric Beneton and his colleagues suggests that the benefits of recreational scuba diving might be greater than those of other sports, decreasing stress levels as well as improving well-being. These benefits may be attributed to divers' slow, full breathing, actions that share similarities with meditation and mindfulness.

The blue mind theory corresponds with the fact that, as mammals, humans spend their early life in amniotic fluid. For humans, being immersed in water genuinely resembles our days in the womb. Immersion in water may satisfy our primordial instinct to regain physical union with the womb and correlates with water being a primarily feminine symbol.

φ φ φ

An essential general comment must be made about theories of evolutionary psychology. These theories advocate for the existence of universal innate cognitive modules in all humans, and the association between ancient peoples' primal environment and lifestyle, and modern humans' psychological biases and mental health. Based on these doctrines, the modules were selected and developed during early human history, when our ancestors were hunter-gatherers. Wilson asserted that genes hold culture on a "leash," meaning culture is not free of genetic constraints. Furthermore, evolutionary psychologists see culture as a thin layer on top of innate, genetically selected, psychological mechanisms. It is not surprising, therefore, that in today's zeitgeist, evolutionary psychological theories are under attack from many directions. They have been criticized for associating disproportionate importance to genetic inheritance, at the expense of social learning and behavioral inheritance. Without a doubt, people were not born as *tabula rasa*.[10] However,

10 *Tabula rasa* refers to the theory that people are born without innate mental content (the mind is a blank slate upon which experience writes) and that knowledge comes from personal experience, perception and socialization.

^ Bennet's (yellow) feather star on a coral reef. Crinoids, which include sea lilies and feather stars, feed on plankton that drift in the currents by filtering the sea water that flows through their feather-like arms. Puerto Galera, Philippines.

many of the phenomena described above could be explained by cultural and social influence (social constructivism). The mainstream debate is not about the existence of genetically innate psychological modules or human traits, but rather about their plasticity and impact on our behavior, compared to socialization and behavioral inheritance.

The feeling of weightlessness

Alongside the relaxation and therapeutic effects of water, diving provides an additional sensation: weightlessness. Weightlessness is a feeling that can be achieved by scuba diving or by using special airplanes that fly on a parabolic orbit so that, for a certain period, they fall freely towards Earth. Weightlessness enables divers to feel like they are flying, controlling their depth by inhaling and exhaling. Jacques Yves Cousteau (1910–1997), a major contributor to recreational diving, described it as follows in a 2003 article in *Time*: "Buoyed by water, he can fly in any direction – up, down, sideways – by merely flipping his hand. Underwater, man becomes an archangel."

Achieving perfect buoyancy is one of the essential skills of a scuba diver and is necessary for achieving the feeling of weightlessness. Neutral buoyancy requires extensive training and special attention to breathing and posture. The idea is to change your diving depth through breath to avoid the use of a buoyancy compensator device (BCD). This attention to breathing, similar to that of meditation, is one of the reasons why divers associate diving with mindfulness. The combination of focused breathing, achieving perfect buoyancy, being immersed in water, and observing nature's beauty makes you feel as if you are entering a different realm.

Intellectual Experience

Many divers focus on the sensual experience of diving, but this is not the sole dimension of diving enjoyment. There is a longstanding debate regarding the contribution of the scientific perspective to the elevation or reduction of appreciating the beauty of the universe. American poet Walt Whitman (1819–1892) was engaged in this debate, but undecided. In his 1867 poem, "When I heard the learn'd astronomer," he expressed his deep disappointment in science's disdain for nature's beauty:

When I heard the learn'd astronomer,
When the proofs, the figures, were ranged in columns before me,
When I was shown the charts and diagrams, to add, divide, and measure them,
When I sitting heard the astronomer where he lectured with much applause in the lecture-room,
How soon unaccountable I became tired and sick,
Till rising and gliding out I wander'd off by myself,
In the mystical moist night-air, and from time to time,
Look'd up in perfect silence at the stars.

In his artistic way, Whitman distinguished between wisdom and knowledge. On the one hand, he perceived wisdom as a process of learning through exploration. On the other hand, knowledge comes from research and scientific theories. This thought could also be associated with what I mentioned previously regarding the limitation of language and speech in expressing sensual experiences.

Unlike Whitman's frustration with science, the diversity of underwater life is, for me, enriched by endless intellectual stimulation. Nobel Prize-winning theoretical physicist and polymath Richard Feynman (1918–1988) described his scientific view as complementary to the perspective of an artist, and not something that hides the beauty of nature behind the details of its parts:

> *I can appreciate the beauty of a flower. At the same time, I see much more about the flower than he sees. I could imagine the cells in there, the complicated actions inside, which also have a beauty. I mean it's not just beauty at this dimension, at one centimeter; there's also beauty at smaller dimensions, the inner structure, also the processes. The fact that the colors in the flower evolved in order to attract insects to pollinate it is interesting; it means that insects can see the color. It adds a question: Does this aesthetic sense also exist in the lower forms? Why is it aesthetic? All kinds of interesting questions which the science knowledge only adds to the excitement, the mystery and the awe of a flower. It only adds. I don't understand how it subtracts.*[11]

This view, that science coincides with the aesthetic experience, is also relevant to diving. My scientific interest in the world of marine creatures does not replace the sensational diving experience; on the contrary, it makes it more complete.

I also believe that scientific thinking itself can ignite emotions of wonder and awe. I have often found myself inspired by universal scientific ideas that connect me to the tapestry of life. One example is the idea that human DNA, a central part of every cell in our body, is directly related to every creature and plant that has ever lived on Earth over the last 4 billion years. Many of our genes are shared with plants and animals, which we often perceive as unrelated to human beings.[12]

Having an interest in underwater life has the advantage of breaking the conventional borders created by our mammalian thinking. Reproduction, mimicry, symbiosis, coloration, and many more dimensions can "open the doors of perception," in the words of Aldous Huxley, demonstrating that there are diverse trajectories

11 https://youtu.be/ZbFM3rn4ldo (retrieved March 6, 2021).

12 Around 60 percent of genes (not the entire DNA) are conserved between humans and fruit flies (and also between humans and bananas). This means that the two organisms share a core set of genes, that may differ in how the genes are expressed.

for evolutionary development and life forms, many of which are surprising and intriguing.

Being an advocate for the importance of the scientific perspective does not mean that I do not recognize its limitations. In the last centuries, science has been the main road for acquiring knowledge, but let's not forget that it is not the only road to acquire wisdom. Science pretends to replace the magical worldview of the ancient world but is driven by the same basic motivation: to understand the world around us. We should never forget that science has not fundamentally promoted our understanding of human consciousness, emotions and experiences; this understanding often comes from non-scientific perspectives. In any event, whether science can completely explain the phenomena of life or not, it can undoubtedly add an explanatory layer that enriches our perspective of life.

˅ Coral pinnacle.
Sinai Peninsula, Egypt.

Reflections on Underwater Photography

You may well ask, "What is the connection between photography and the diving experience?" The straightforward answer is that, from an early stage, photography was an integral part of my diving experience. On a deeper level, photography represents for me the tension between the sensual and the intellectual, between the perspective of the observer and the participant. In his bestselling book *Thinking, Fast and Slow*, Nobel Prize winner Daniel Kahneman (1943–) presented the concept of two conflicting selves, the experiencing self and the remembering self, a distinction that has a lot to do with the idea and popularity of photography in modern life. In Kahneman's words:

> *The photographer does not view the scene as a moment to be savored but as a future memory to be designed. Pictures may be useful to the remembering self – though we rarely look at them for very long, or as often as we expected, or even at all – but picture taking is not necessarily the best way for the tourist's experiencing self to enjoy a view.*

My relationship with photography has always been conflicted. On the one hand, photography allows us to "confiscate" a particular event from the constant flow of time and make it stationary, unlike nature's endless flux. This is what gives photography its central uniqueness. On the other hand, I was always deterred by the potential of photography to distance the photographer from the subject, causing the photographer to adopt an observer's perspective, rather than becoming an emotionally involved participant. American intellectual Susan Sontag (1933–2004), in her book, *On Photography*, referred to the relationship between images and consumer society: "Cameras miniaturize experience, transform history into spectacle. As much as they create sympathy, photographs cut sympathy, distance the emotions."

My diving experience has been highly influenced by these principles. At the beginning, I tried to perpetuate the sublime experience, but learned that it is very challenging to replicate such awe and exhilaration. The photographer's ability to create a photo that "stings" the viewer and breaks the general common picture, which is interesting but without exceptional acuity, is challenging to achieve in underwater photography.[13]

13 In the words of French essayist Ronald Barthes, "it is this element which rises from the scene, shoots out of it like an arrow, and pierces me" (Barthes, 1981).

Often, I decided to put my camera aside and focus on the present moment, the here and now, the sensational experience of the dive. I did this because I believe that the singularity of the present moment is all we really have (temporally). The feeling of weightlessness, the rich colors, the awe-inspiring scenery, the ability to get close to and feel part of nature – all of these were an integral part of my diving experience. I did not allow the observer attitude to take over my experience, and I often tried to participate in the experience without thinking, "How will it look when I come back home?" Resolving those two contradicting selves has become a central goal for me.

Another angle that sheds light on the photographic experience can be derived from the observation by French philosopher Alain Badiou (1937–), a point later elaborated by Slovenian philosopher Slavoj Žižek (1949–), that a key feature of the twentieth century is "passion for the real." Modern times are characterized by artificial products that provide reality devoid of substance: coffee without caffeine, beer without alcohol. In our hyperreal world, we are unable to distinguish between reality and the simulation of reality that surrounds us (simulacra, according to French theorist Jean Baudrillard).[14] This virtualization process, which sometimes detaches from the original referent, causes us to experience virtual reality as reality, sometimes more "real than the real." Virtualization also creates a passion for experiencing "the real" in a non-instrumental manner. Entertainment, information, and communication are platforms that provide this hyperreal experience and influence people's thoughts and behavior.

On the one hand, the instrumental encounters with real nature seem to fulfill our desire to get out of the hyperreal environment. Diving provides an opportunity to escape from the hyperreal world. On the other hand, underwater photography is an enjoyable hobby that is also part of the mechanism that structures the realm of hyperreality and distances us from nature.

Hunting instinct

The last point I want to mention is the connection between animal photography and hunting. Throughout the first half of the twentieth century, scientists spoke of a hunting instinct, arguing that violence is deeply embedded in human nature, citing

14 In his book *Simulacra and Simulation* (Baudrillard, 1994), Baudrillard claims that in the media and in consumer society, people are caught up in the play of images, spectacles and simulacra, that they have less and less relationship to an outside, external reality, to such an extent that the very concepts of the social, political, or even reality no longer seem to have any meaning (https://plato.stanford.edu/entries/baudrillard/).

evidence that humans hunted as early as two million years ago. It is a controversial idea and no longer popular. Most of the scientific community believes that our larger brains evolved mainly to help us cooperate. In the last centuries, hunting animals that are not submissive to human authority has become a means of displaying the power and dominance of humanity over nature. High-stakes trophy hunting, both in terms of money and danger, probably serves as an expensive method of demonstrating the hunters' fitness to rivals and potential mates and is not comparable to the activity of real hunters who hunt for food. Whether hunting is hard-wired, culturally acquired, or a "just-so story,"[15] many people enjoy animal photography in the same manner that they may enjoy hunting. The advantage of photography, of course, is in its lower negative impact on nature.

« Colorful coral reef. Verde Island, Philippines.

Alongside the drawbacks of the photographic experience, underwater photography is a challenging, intriguing and enriching hobby. It provides me with a powerful tool that conveys my visual underwater experience. As time goes by, you become familiar with the names of the animals, their behavior, natural habitats, and unique characteristics. Spotting familiar creatures becomes an ongoing challenge. It is not just about knowing where to find them; it also relates to the ability of your vision system to quickly interpret the hidden patterns of small, camouflaged creatures.

15 A just-so story is a speculative, unprovable story or explanation for a cultural practice, biological phenomena or human behavior.

"Passing up the harbour, in appearance like a fine river, the clearness of the water afforded me one of the most astonishing and beautiful sights I have ever beheld. The bottom was absolutely hidden by a continuous series of corals, sponges, actinia, and other marine productions, of magnificent dimensions, varied forms, and brilliant colours. ... In and out among them moved numbers of blue and red and yellow fishes, spotted and banded and striped in the most striking manner, while great orange or rosy transparent medusa floated along near the surface. It was a sight to gaze at for hours, and no description can do justice to its surpassing beauty and interest."

CHAPTER 2

Coral Reefs
A Holistic View

« A school of scalefin anthias swim around a brain coral. Like many coral species, brain coral colonies grow via asexual reproduction known as budding. Eilat, Israel.

Coral reefs were known to ancient cultures of harvesting divers, as well as to the mariners of the ancient literate civilizations of Mesopotamia and Egypt. Attention to them was mainly due to the hazard they posed to navigation in the Red Sea and Indian Ocean. Likewise, during the era of imperial expansion, European explorers that rounded the Cape of Good Hope on their journey to the Indian and Pacific Oceans encountered coral reefs that sometimes severely damaged their ships. One of the most famous stories of encountering coral reefs is the accidental discovery of Australia's Great Barrier Reef by Captain James Cook (1728–1779). On June 11, 1770, 10:30 a.m., Cook's ship, HMS *Endeavour* was severely damaged when it ran aground on a shoal later described by Cook as an "insane labyrinth" of coral. Desperate to lighten the ship, the crew threw overboard nearly 50 tons of cargo. Twenty-three hours later, during the next high tide, the *Endeavour* pulled free and managed to reach a safe harbor. The crew spent a few weeks repairing the ship before it could return to its original mission. At the time, mariners such as Captain Cook had no idea of the complex science behind the reef's biological formation.

Many shipwrecks during the eighteenth century ignited scientific interest in the nature and mechanism behind coral reef habitat formation. However, it was only during the nineteenth century, following the voyages of Captain Matthew Flinders (1774–1814), an English cartographer who explored the Australian coastline, and Charles Darwin (1809–1882), that people started coming up with theories about the science behind the formation of coral reefs. Darwin understood that atolls are produced by biological activity that creates the coral reef system around volcano craters.[16] His

16 An atoll is a coral reef that partially or completely encircles a lagoon. Atolls commonly rest on the peak of sub-marine volcanic structures.

book about atoll formation was published in 1842, long before his theory of natural selection was published in 1859. Furthermore, the romantic aesthetic perception of the sublime, which took hold in the early nineteenth century, inspired many scientists and explorers. One of them was British naturalist Alfred Russel Wallace (1823–1919), the lesser-known but equally significant co-founder of the theory of evolution, who described his amazement at coral reefs in his book, *The Malay Archipelago*.

φ φ φ

On an evolutionary timescale, human interest in coral reefs, whether from a nautical or scientific perspective, is a mere blink of an eye. Since the emergence of life on Earth around 3.5 billion years ago, reefs in some shape or form (not necessarily coral reefs) have existed, making them one of the oldest ecosystems on the planet. For about 3 billion years, the majority of life forms were, at most, simple colonies of individual cells, or simple multicellular creatures. These microbial communities, that produced hard substrates, creating rocky structures, are known today as the oldest reef-building organisms.

The Cambrian explosion, around 542 million years ago, marked the "Big Bang" of biological diversity and a significant change in the variety, diversification and complexity of life forms. The phyla of many of today's marine animals originated and diversified during the Cambrian evolutionary radiation.[17] Only at a later stage, and following several significant mass extinctions, did plants and animals invade terrestrial areas (200–400 million years ago). This is one of the reasons behind the higher-level taxonomic diversity in marine environments compared to terrestrial ones. The long separation of evolutionary trajectories among marine lineages has resulted in a greater variety of body plans, functional and biochemical diversity, and the subsequent emergence of more species.

Coral reefs as we know them today are based on stony corals that appeared more than 220 million years ago. Of the thousands of species that inhabit coral reefs, only a fraction generate the rigid limestone skeleton that builds the reef.[18] Evidence suggests that stony corals formed symbioses with algae soon after their emergence, an adaptive event that allowed for the proliferation of coral reefs.

---

17 Phyla (singular phylum) is a biological taxon between kingdom and class.

18 There were also older reef structures made by the ancestors of modern corals. Nonetheless, *Scleractinia* (the dominant modern reef-building corals), as a group, is estimated to have originated about 160–220 million years ago.

‹ Reefs are composed of colonies of stony coral polyps that serve as a platform for a wide variety of other animals. They are home to corals and other filter feeders, including sponges, ascidians and feather stars. The idea that reefs are made up of animals surprises many people, who tend to think of reefs as colorful plants or rocks. In fact, a reef can be more accurately imagined as a "surface of tiny mouths." Verde Island, Philippines.

The diversity of corals, colored fishes and invertebrates, mostly along the equator but also in subtropical areas, represents a long and remarkable evolutionary history. When reviewing the data behind coral reefs, one cannot overstate the uniqueness and importance of this habitat.

Similar to terrestrial rain forests, coral reefs are isolated areas with extreme biodiversity. For example, 32 of the 35 known animal phyla are found in coral reefs, compared to only nine phyla in tropical rainforests. The average number of species that live in coral reefs is roughly estimated to be around 830,000 multicellular species; about 80 percent of them have not been discovered yet (a true *mare incognita*!).[19] The estimated number of coral reef fish species is more than 5,000, around 10–20 percent of all marine fish species.[20] In the Coral Triangle alone, an area which extends from Indonesia to the Philippines, there are more than 4,400 marine fish species.

In terms of size, coral reefs occupy a relatively small area, about half the size of Madagascar. This accounts for less than 1.2 percent of the world's continental shelf, and only about 0.1 percent of the total area of the world's oceans. Nevertheless, Australia's Great Barrier Reef, considered to be the world's largest living organism structure, can be seen from outer space. This large spine of life stretches for more than 2,200 km, approximately the distance from Brussels to Moscow.

19 According to a 2015 study by Fisher et al.

20 Based on Fishbase, there are about 34,300 total fish species: https://www.fishbase.de/search.php (retrieved March 6, 2021).

> A camouflaged scorpionfish surrounded by sponges. Bunaken, Indonesia.

˅ Moorish idols and royal angelfish. Bunaken, Indonesia.

< Magnificent coral reef. Verde Island, Philippines.

What Are Corals?

The central narrative of this book evolves around relatively simple animals: corals. Corals are two-layered invertebrates characterized by radial symmetry. In terms of taxonomy, they are related to jellyfishes and sea anemones, which are all part of the various classes within the phylum Cnidaria. Most reef-building corals live in the shallow waters between the latitudes of 30°N and 30°S; the waters are crystal clear, warm and low in nutrients, with an oceanic salinity.

Corals are made up of tiny individual animals called polyps. Each polyp's mouth is symmetrically surrounded by a ring of stinging tentacles that enable it to capture floating prey (mainly plankton).[21] In most cases, the tentacles are arranged with six- or eightfold symmetry. The prey is transferred into the polyps' mouths and digested in their digestive sac. In addition to actively catching floating prey with their hunting arms (tentacles), most corals also have an external mucus layer, that is used to deliver the prey, dissolved nutrients and inorganic matter from the water to the digestive cavity. Most corals feed at night, benefiting from the vertical migration of their prey (zooplankton) from deep water to the water surface at sunset.

21 Plankton are a diverse collection of organisms with limited swimming capabilities that drift with the current and live in large bodies of water.

Typical anatomy of a stony coral

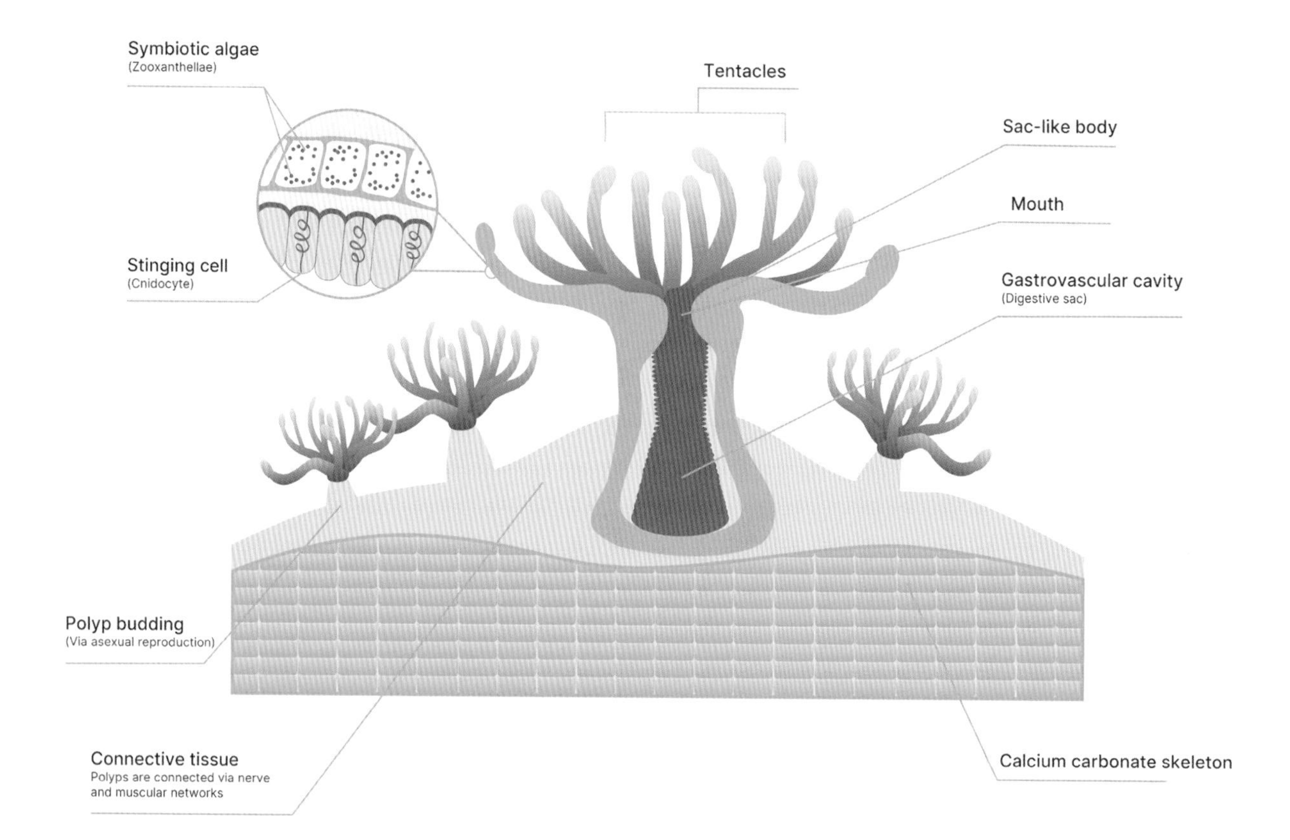

For simplicity, we can divide corals into two major types: stony corals (Scleractinian) and soft corals. Stony corals, billions of which form the coral reefs, extract calcium carbonate (the same material as is used in the formation of many seashells) from seawater to build their limestone skeleton. Soft corals, as suggested by their name, do not produce such a skeleton. Instead, their bodies contain tiny, needle-like structures called spicules, made from hardened calcium particles, that provide them with the required structural support.

Coral as an animal

One of the most striking facts about corals is that they are animals, specifically sessile animals, meaning they are largely immobile. It is not surprising, therefore, that until late antiquity, corals were usually defined as plants or minerals, or something in between. Aristotle (384–322 BCE) classified corals as plant-animals (zoophytes).

Interest in corals in antiquity was primarily confined to jewelry-making. An example is red coral (*Corallium*), used in ancient Rome as an amulet to protect against the evil eye and bad spirits.

Perhaps the first person who argued that corals and sponges should be considered animals was eleventh-century Persian polymath Al-Biruni (973–1050), whose argument was based on how corals react when touched; but for many years later, coral was still considered by most scholars to be a plant or mineral. In the sixteenth century, French naturalist Petrus Gyllius (1490–1555) defined coral as a "third nature" hybrid plant and animal. After the invention of the microscope, and eight hundred years after Al-Biruni, another great polymath, William Herschel (1738–1822), could substantiate that corals are, indeed, animals. Herschel looked at coral under a microscope and realized that coral cells lacked cell walls, a distinct non-animal structural feature that surrounds the cells of plants and algae.

Another reason why corals are part of the animal kingdom, in addition to the characteristics described above, is that corals, like animals, cannot produce their own food. This differs from plants, which can produce their own, and are therefore known as "primary producers." Corals are heterotrophic, meaning they consume nutrients from other organic sources. It is true that, like plants, most corals are sessile animals;[22] however, in their early phase of development, as larvae,[23] they are free-swimming (locomotion is another characteristic of animals) and drift in the ocean currents with plankton. Corals also have another unique characteristic of animals: they have a nervous system, albeit a very simple one.

Coral reproduction

Corals exhibit a diverse range of reproductive strategies, and these are among the most important features contributing to their long-term ecological and evolutionary success. Coral sexuality ranges from species that have separate sexes (gonochorism), meaning that every individual colony is either male or female, to species in which every coral serves as both male and female (hermaphroditism) and produces both sperm and eggs simultaneously. Some corals can even change their sex from one

22 There are some exceptions. For example, the mushroom coral is a solitary species and capable of benthic locomotion. Many sessile creatures are characterized by radial symmetry, an efficient morphology that can sense the environment and collect food from all directions.

23 Being a larva is a developmental stage in the life cycle of certain animals that juveniles undergo before metamorphosing into adults. This stage is characterized by various structures and organs that change while the animal transforms into an adult.

reproductive season to the next, serving as males one season and females in the next season, or vice versa.

Most corals are attached to the seafloor and cannot move. So how do they find a mate in the ocean? The majority of reef-building corals reproduce sexually through broadcast spawning, which is the spawning of their reproductive material (eggs and sperm) into the open sea, where external fertilization takes place. Since both eggs and sperm remain viable for only a few hours in the water and are also, within a short period, diluted by the ocean currents, the timing of the spawning event is essential for its success. Successful fertilization, which can only take place during this narrow time frame, has led to the evolutionary trait of precise spawning synchrony orchestrated by environmental factors. For example, the precise month of spawning might correlate with seasonal changes in solar radiation, sea temperatures and the wind. Lunar cycles might set the exact night of the month, while sunset might cue the precise hour.

Coral spawning, described as the greatest orgy in the world, presents one of nature's most astonishing examples of synchronized phenomena. It usually happens only once a year and takes just a few minutes. Within seconds, as if a mysterious natural force had turned on a hidden switch, a huge number of corals of a given species spanning vast areas simultaneously spawn their reproductive material in what looks like a colorful underwater snowstorm. The spawned eggs and sperm are carried away in the currents until they join, the sperm fertilizes the eggs, and these develop further and create planktonic larvae called planulae. That this magnificent phenomenon is under threat was proposed by Shlesinger and Loya in their 2019 article "Breakdown in spawning synchrony: A silent threat to coral persistence." They suggest that this extraordinary synchronization of spawning may be violated by human and natural disturbances, which might pose a severe hidden threat to successful coral fertilization.

Another mode of sexual reproduction employed by certain coral species is brooding. In this mode, corals release sperm, which reaches other corals, and the fertilization and planulae development take place internally, within the coral polyps. These planulae will later be released as polyps into the ocean, navigating via chemical cues, light and possibly reef sounds through the water until they settle on the hard reef substratum and metamorphose into a primary polyp, the very beginning of a juvenile coral. Then, they may divide or bud off,[24] forming a series of interconnected polyps, creating a coral colony that could live for dozens or even hundreds of years.

24 A form of asexual reproduction in which each individual uses only its own DNA to create offspring, rather than DNA from two individuals, as in sexual reproduction.

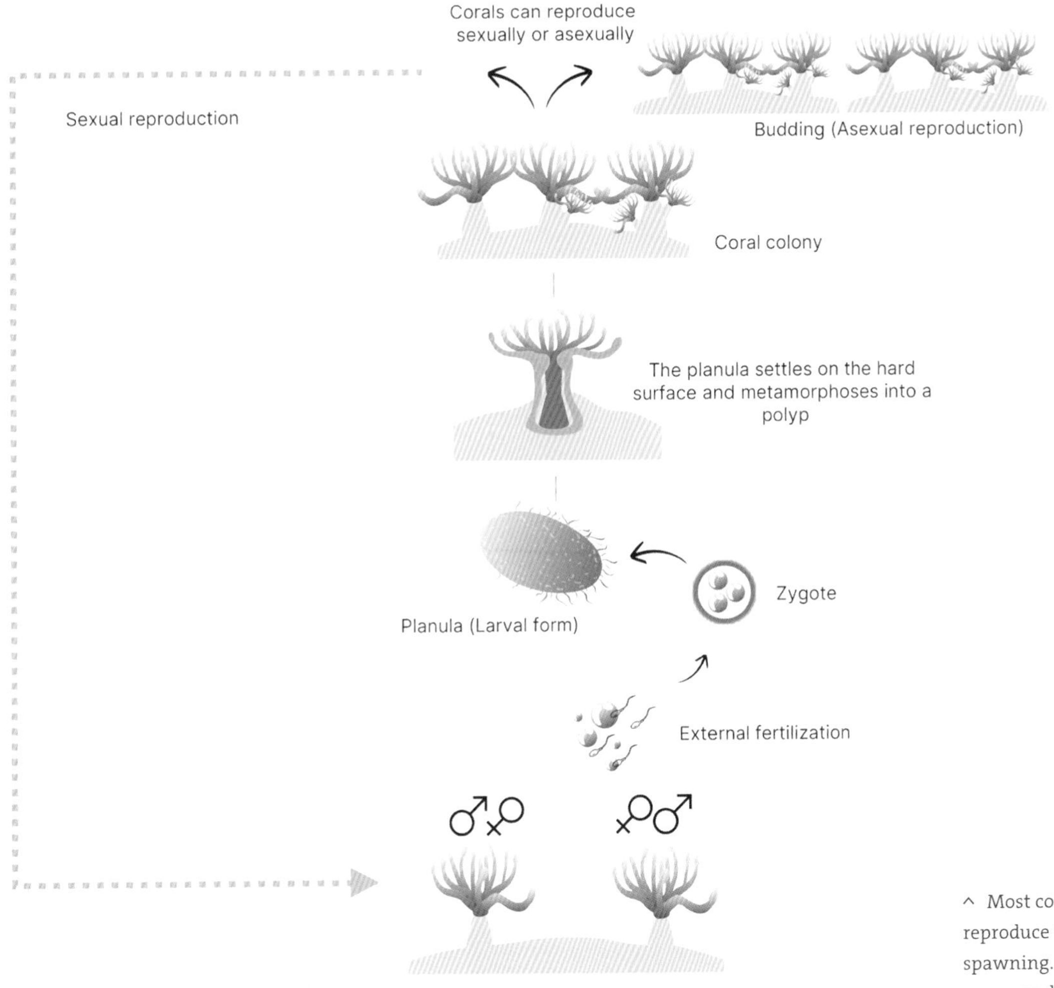

^ Most coral species reproduce via broadcast spawning. They produce both sperm and eggs at the same time and the fertilized eggs, known as planula larvae, swim at the ocean surface for days or weeks, eventually falling back to the seabed and attaching themselves to a hard surface. The larvae then metamorphosize into polyps, begin to grow by budding (asexual reproduction) and form a coral colony. As the coral colony grows and matures, it begins to sexually reproduce through broadcast spawning.

Polyps within a colony share a skeleton and are linked by living tissues through which they share and circulate nutrients and other materials. Once a colony reaches a certain size or age, it attains sexual maturity and starts reproducing sexually.

The fact that coral reproduction is carried out in two ways, sexual and asexual, might be an essential condition to the future persistence of coral reefs. These two avenues have the potential of increasing the resilience and adaptation of coral reefs to environmental changes since beneficial mutations may take place in either avenue. In other words, in addition to sexual reproduction, introduces large genetic diversity but is characterized by long generation time (years), beneficial mutations in somatic cells can also, over time, introduce genetic innovations, and thus may also improve corals' fitness and resilience in rapidly changing environments.[25]

Symbiosis with algae

The polyp's symbiosis with algae is one of the key factors in understanding the proliferation and radiation of the coral reef system. Most stony coral polyps live in an obligate mutualistic symbiotic relationship with the photosynthesizing microalgae (zooxanthellae) that live within the coral tissue.[26] One cannot overstate the importance of this symbiosis as a cornerstone of coral reef development; one which accounts for their evolutionary success. This type of association, between single-celled algae and coral, is known as photosymbiosis, and it also exists in other marine invertebrates including jellyfishes, sea anemones, bivalves, sponges, nudibranchs, flatworms and hydra.

While previously thought to have developed about 60 million years ago, recent genetic studies suggest that the symbiotic relationship between reef-building corals and algae is even more ancient and may have emerged as early as 160 million years ago. This timing corresponds with the evolutionary radiation of reef-building corals and might indicate reef algae resilience, as they have managed to survive mass extinctions, including that which wiped out most dinosaurs.

>> This close-up view of a soft coral from the Xeniidae family shows individual polyps with a distinctive soft-coral feature: each polyp has eight large feathery tentacles. Eilat, Israel.

> A green anchor (hammer coral) (*Euphyllia*). Although this coral resembles a soft coral, it is actually a stony coral in which the large, fleshy tentacles hide a hard skeleton underneath. Lembeh Strait, Indonesia.

25 Somatic cells are any cells in the body that are not sperm or eggs (gamete).

26 Zooxanthellae is a collective term for single-cell photosynthetic organisms that are part of the phylum Dinoflagellate.

« Orange cup or sun coral (*Tubastraea* sp.), with extracted tentacles at night. Many stony corals retract their polyps during the day when predators are active and abundant. Lembeh Strait, Indonesia.

‹ Fan corals (*Gorgonia* sp.) are soft corals that typically grow perpendicular to the water current. Some fan corals contain symbiotic algae, which provide nutrients through the byproducts of photosynthesis. Verde Island, Philippines.

Algae use sunlight for photosynthesis and transfer 95 percent of the food they produce to their coral host.[27] In return, the algae receive protection from currents and predators, as well as some nutrients. For corals living in areas viewed as "oceanic nutrient deserts," this symbiosis provides stony corals with sufficient energy to generate the limestone skeletons that form the tangled structure of reefs, providing shelter and habitat for a considerable number of species. Since corals both actively feed on plankton and have a mutualistic symbiosis with photosynthetic algae growing within their tissues, they can be viewed as both "farmers" and "hunters."

Due to the dependency of many corals on the food produced through photosynthesis, most reef-building corals are restricted to a relatively specific range of temperatures and depths, where sunlight still penetrates (this can be more than 100 meters deep in some locations). This relatively shallow-water habitat is naturally

‹ While most stony coral species live as colonies of polyps, some species, such as this mushroom coral from the Fungliidae family, live as a solitary, free-living, single polyp. Puerto Galera, Philippines.

27 In a nutshell, photosynthesis converts light energy into chemical energy by converting carbon dioxide, water and light into sugars and oxygen ($6CO_2 + 6H_2O$ + photons (sunlight)$\rightarrow C_6H_{12}O_6 + 6O_2$). It is one of life's most extraordinary and fundamental processes. This release of oxygen also played a significant role in generating the Cambrian explosion, and the evolution of complex and conscious living systems.

›› A bubble coral shrimp among bubble coral (genus *Plerogyra*) polyps. This shrimp is considered to be an obligate commensal and is found only on this type of coral. Lembeh Strait, Indonesia.

› Elephant skin coral (*Pachyseris* sp.). Many algae-possessing corals are characterized by a greenish-brownish color. During the day, when sunlight is abundant, these symbiotic algae provide their coral host with food produced during photosynthesis. In return, the algae receive nutrients that they need, and protection. Puerto Galera, Philippines.

› Close-up view of massive coral polyps. Lembeh Strait, Indonesia.

more vulnerable to climate change than deep water. The global warming catastrophe has disrupted the delicate symbiosis upon which many corals depend. The stress caused by temperature anomalies, in combination with high levels of light, results in the breakdown of the coral–algal symbiosis. During these stress events, corals eject the symbiotic algae living in their tissue, a process known as coral bleaching. This leaves their tissue transparent, so their white calcium carbonate skeleton is visible. Many corals are dependent on these algae for their energy, so without it, the corals are subject to starvation, more prone to disease, and may eventually die.

Biodiversity in the Coral Triangle

Anyone who compares the experience of diving in the Red Sea or the Caribbean to diving in the Philippines or Indonesia immediately notices a difference in biodiversity (see map below).[28] When diving in places such as the Bunaken Marine Reserve or the Lembeh Strait, both in Indonesia, you may well encounter several frogfishes or octopuses during one dive. In the Caribbean, you are lucky to meet one. It is common to see local dive guides in the Philippines and Indonesia, many of whom are gifted with fantastic spotting skills, getting out of the water with more than 20 noteworthy species of fishes or invertebrates written on their underwater dive slates. From the perspective of a lay diver, someone who is not a marine biologist, biodiversity usually means two things: the variety of species in a specific area and the frequency of encountering them. While diving in the Red Sea, I was lucky to encounter many species, but the occurrence of encountering them was very low. The reason behind those differences has been on my mind for many years.

Global biodiversity of reef-building corals

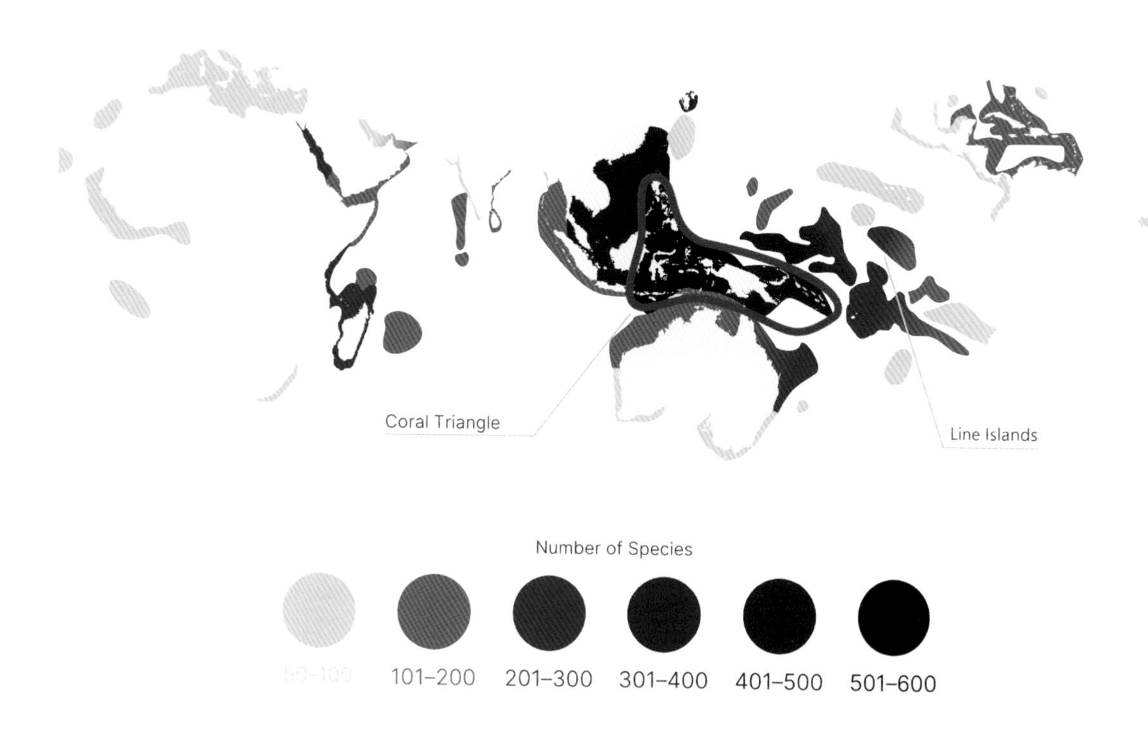

28 The biodiversity map is based on Veron et al., 2009; Veron et al., 2015 and Bruke et al., 2011.

For more than 300 years, Western science has known that biodiversity is greatest in the tropics; that is, biodiversity increases as one moves from the poles to the equator. Known as the latitudinal diversity gradient (LDG), this occurs in the vast majority of flora, fauna, and fungi, and takes place at all levels of evolutionary differentiation: species, genetic, phenotypic, etc.[29]

Possible explanations for LDG can be divided into four major, non-exclusive theories. The first theory asserts that during glacial periods, living conditions in the tropics were much more stable and hospitable than those in higher latitudes, enabling a long evolutionary process and higher degree of speciation. The second theory attributes the exceptional biodiversity to higher temperatures and humidity, which provide better nurturing conditions for many organisms. A third theory suggests that the year-round proliferation of parasites and pests in the tropics might limit the growth of populations, preventing any single species from dominating the habitat and enabling the emergence of species diversity in a specific geographical location. The fourth theory is attributed to higher productivity in the tropics (productivity being defined as the amount of biomass generated per specific period), resulting from abundant solar energy. This abundant solar energy means greater "effective" evolutionary time: faster metabolic rates, shorter generation times, faster mutation rates and faster selection.

There is another significant biodiversity gradient along the equator, one which mainly relates to marine creatures and coral reef habitats. This biodiversity is most pronounced in the Indian Ocean and western Pacific Ocean, and decreases with increasing distance from the Indo-Australian archipelago. While the pattern of LDG is well known, the reason for the longitudinal marine gradient (LMG) remains poorly understood.

The Coral Triangle is the global center and apex of this longitudinal marine biodiversity gradient. The triangle sprawls between the Philippines in the north, Indonesia and Malaysia in the west and southwest, and Papua New Guinea and the Solomon Islands in the southeast. The diversity of the Coral Triangle is incredible, perhaps even rivaling the Amazonian rainforests and indeed any other terrestrial biodiversity hotspot.

Compared to other terrestrial systems, the Coral Triangle exhibits significant east-to-west longitudinal biodiversity gradients. However, unlike terrestrial areas, the biodiversity of the Coral Triangle is high not just compared to areas with different

---

29 There are some exceptions, for example penguins, where diversity increases towards the poles.

latitudes, but also compared to areas along the equator (different longitudes). An illustration of the significance of the Coral Triangle gradient can be found in the number of fish species in the Philippines, which may exceed 2,000, compared to the numbers of fish species in Hawaii, the Bahamas, and the Mediterranean, which are 450, 500 and 250, respectively. This striking, non-intuitive biodiversity gradient raises the question: "Why are there so many species in the Coral Triangle?"

Four theories offer explanations for this richness. The first is the center of origin theory, which asserts that more new species originated in the Coral Triangle and radiated towards other areas from this epicenter. Based on this theory, in some cases, ocean currents may play a significant role in limiting the expansion of marine species to other areas. The second is the center of accumulation theory; it postulates that species originating in other places preferred, or were forced by geographical factors such as ocean currents or island collision, to settle in the Coral Triangle. This theory is essentially the opposite of the first. Another possibility is the center of overlap theory, which defines the Coral Triangle as a center where sister species in the Indian and Pacific Oceans overlap, creating elevated biodiversity. The fourth is the center of survival theory, which suggests that a low rate of extinction, compared to other areas, characterizes the Coral Triangle. This theory asserts that multiple species have managed to remain within the center, regardless of their geographic origins (survival in refuge). This survival is the result of more habitats, including deep areas, larger total reef surface area, and larger population sizes. These conditions might evolve in a positive feedback loop; that is, as the reef increases, more and more species find refuge in different niches.[30]

The common challenge of all of these theories is the question of where speciation took place.[31] Determining this is difficult because most marine creatures have a dispersive pelagic larval stage, during which they can travel great distances. In some coral species, for example, laboratory experiments show that planulae can live for a significant period of time, even a few months, before settling and forming a colony. Theoretically, then, they could be transported long distances by ocean currents. Due to this pelagic long-distance dispersal character, coral reef biodiversity has a

30 Russian biologist Georgy Gause (1910–1986) suggested the competitive exclusion principle, according to which the number of species in any environment can be no more than the number of available niches; that is, there must be a match between species and their role and position in the environment. A different way to phrase this principle is that when resources are limited, complete competitors cannot live together, since when one species has even the tiniest advantage over the other, it will eventually dominate. This applies to coral reefs, but has been challenged by examples that refute it; for instance, plankton live in environments with limited niches but their diversity is enormous.

31 Speciation is an evolutionary process by which populations diverge and form distinct new species.

^ A giant barrel sponge. Sponges, unlike most multicellular organisms, are characterized by asymmetrical body plans. They are among the world's most ancient multicellular animals and started evolving about 635–680 million years ago, prior to the Cambrian explosion. Verde Island, Philippines.

special significance in the evolution and radiation of Earth's biota.[32] Since the pivotal moment of the Cambrian explosion, tropical reef environments have been a source of biodiversity, hence the name "evolutionary cradle." Moreover, tropical reefs are more than evolutionary cradles; they are also responsible for exporting biodiversity outwards.

Another approach towards resolving the LDG mystery was suggested by phylogenetic methods and takes into consideration factors such as the colonization time and diversification rates (speciation minus extinction). The unique diversity of the Coral Triangle can be explained by the colonization of many lineages 5 million to 34 million years ago. Those colonizations allowed more time to accrue, speciate, and create species richness compared to other tropical marine regions. Thus, diversity in the Coral Triangle was mainly due to a few ancient colonizations that were characterized by high diversification rates. These results support the center of accumulation and center of overlap theories, and their main conclusion is that species richness is a long-term process; if damaged by humans, recovery could take millions of years.

32 Although larvae have a sufficient amount of energy to sustain long-distance dispersal, factors such as currents and environmental gradients can significantly limit dispersal distance.

« Various types of ascidians (sea squirts). Puerto Galera, Philippines.

What Does a Pristine Coral Reef Look Like?

The impact of modern human culture on coral reefs is devastating and has worsened in recent decades. But it would be wrong to attribute this destruction solely to the modern period. Humans' impact on nature has been a fact of life ever since humans developed "advanced" consciousness and an anthropocentric attitude towards the world. While it is challenging to pinpoint when this cognitive revolution took place, researchers agree that a significant leap occurred when fictive language and advanced culture evolved around 70,000 years ago.

The arrival of humans in Australia around 50,000 years ago is what caused the mass extinction of numerous species of large mammals that had lived there for thousands of years. Likewise, the high extinction rates of large carnivores in East Africa is associated with increased hominid brain size and changes to vegetation.[33]

Over the past 10,000 years, during the period known as the Anthropocene,[34] the far-reaching impact of humans on nature has grown; this is particularly true over the last two centuries. The gradual process of disengagement from nature, combined with ever-increasing domination over the natural world, accelerated during the Age of Enlightenment. Unlike terrestrial areas, and due to the natural barrier of the sea, coral reefs were relatively immune to human impact, on an evolutionary scale, until recently, and their significant ecological degradation only began with the rise of colonial occupation, around the seventeenth century. This is earlier than most people think, and represents a significant amount of time in terms of potential accumulated damage. This also raises the question, "What did pristine coral reefs look like before human disturbance?"

φ φ φ

Diverse metrics and indices have been developed to evaluate coral reef health. These include coral and macroalgae benthic coverage, reef species richness, diversity, evenness and biomass, the type of microbial community, and more. One metric that represents a specific point of view is the biomass pyramid, a diagram that describes a population's biomass at each level of the food chain. At the bottom of the pyramid are producers, above them are the primary consumers, and so on, up to apex predators such as sharks. This point of view yields some intriguing results.

33 The extinction rate of small carnivores did not change.

34 The Anthropocene is the most recent period in Earth's history, marked by the start of humans' significant impact on Earth's geology and ecosystems.

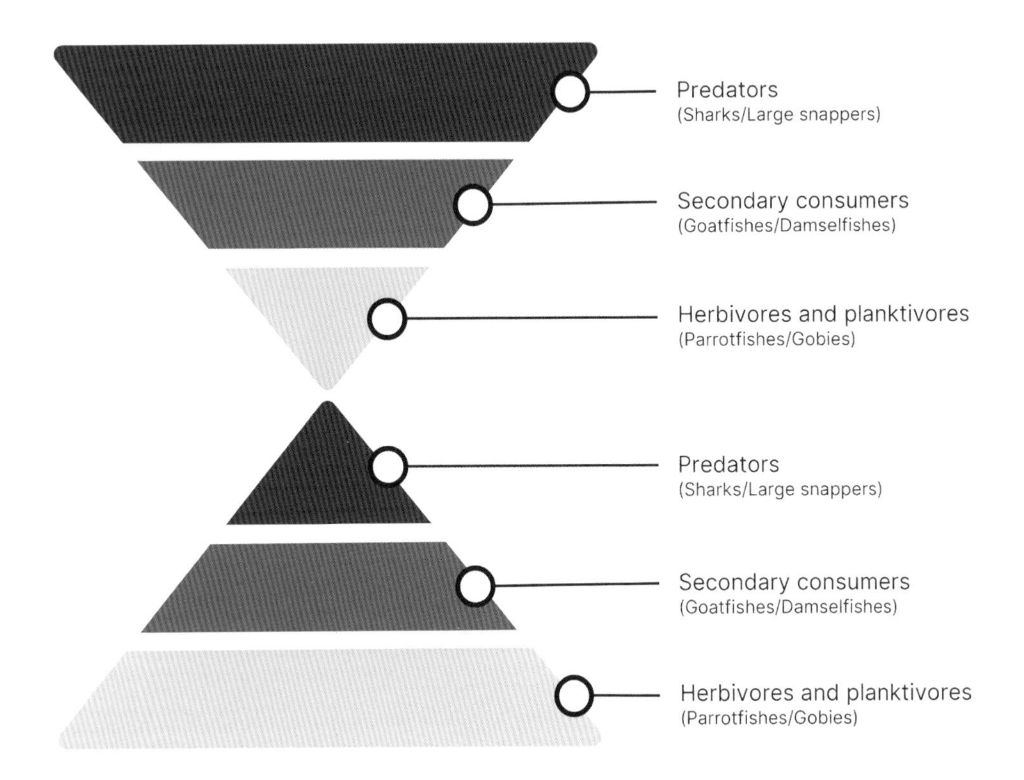

For many years, the prevailing paradigm assumed that most of a healthy reef's biomass is concentrated at the bottom of the pyramid, with herbivores that feed on algae, and above them, secondary feeders and large predators. However, a survey conducted in Kingman Reef,[35] considered by some scientists to be an unexploited, near-pristine reef, concluded that the biomass pyramid of today's coral reefs might have a different structure than the pristine reefs of the past, before pronounced human intervention. Researchers suggested that a pristine reef's spatial biomass structure is spread in an inverse pyramid pattern, where 85 percent of the biomass is attributed to large predators. This contrasts with most reefs today, which suffer from overfishing that tends to remove large, slow-growing animals first, and gradually, the

35 The uninhabited and isolated Kingman Reef and Palmyra Atolls are part of the Northern Line Islands, located in the North Pacific Ocean. These reefs are about 5,400 km from the nearest continent and are assumed to provide a baseline for the natural state of coral reefs.

smaller fish. Furthermore, a model based on data from Kingman Reef and Palmyra Atoll, both located in the Line Islands archipelago, suggests that high fishing pressure might destroy the inverted pyramid. This model also shows that the fishing of large predators alone will have this effect. In other words, what we consider to be pristine might actually be a consequence of biased modern observations that took place long after humans had already started to exploit apex predators in the coral reef ecosystems.

These results raise the question of how 85 percent of the biomass could survive on only 15 percent of the biomass. The answer lies in creatures' life cycles. Animals at the lower parts of the food chain (the inverted pyramid) reproduce and grow very quickly. In the upper part, we find animals that live for many years and reproduce very slowly. In other words, the production surplus of creatures at the bottom is enough to maintain the biomass of those at the top.

Over the past few years, the hypothesis of an inverted trophic biomass pyramid, which was never widely accepted, has been challenged by research which found that it inaccurately represents the overall biomass structure of coral reef ecosystems.[36] New results regarding the density of sharks in Palmyra Atoll, obtained through the use of advanced measurement methods, have shown that shark density is significantly lower than predicted by the inverted trophic biomass pyramid hypothesis. In other words, the jury is still out on this one, and the trophic structure of an unexploited reef might not be inverted at all. A potential positive aspect of this research is that the recovery of coral reefs might be a much more realistic mission than previously thought, since the target number of large predators is lower than that predicted by the inverted pyramid hypothesis.

Reef sounds

For divers, reefs are perceived as silent places, in part because of the noise made by diving equipment. However, there is actually an acoustic cacophony in the reef, produced by organisms in underwater environments. This was first identified during the Second World War, following the invention of sonar. New technologies enabled the development of novel methods for assessing some aspects of reef health by listening to reef ambient sounds. These sounds are produced by a wide variety of marine creatures and play a

36 The research asserts that the number of fast-moving animals such as sharks was over-counted, and that the population of slow-moving, benthic animals was more likely to have been accurately assessed.

significant role in the functioning of marine communities.[37] Even seaweed and algae produce sounds while creating gas bubbles during the photosynthesis process. Sound is used as a signaling method for reproduction and feeding, and is a by-product of creatures' activities. The snapping shrimp, for example, also known as the pistol shrimp because of its asymmetrical claws, generates sounds for hunting as well as for communication.[38]

Even coral larvae might use sound to locate a suitable benthic settlement area.[39] Unlike cues from chemical compounds produced by reef organisms that can be detected by larvae when they come into close proximity to the reef, and that are also dependent on the current's direction, reef sounds may provide an orientation cue for more considerable distances.

Since many fish species produce sound, the diversity of sound types may be indicative of fish assemblages and provide information about the state of the reef. Furthermore, scientists have recently demonstrated that playbacks of healthy reef sounds can increase local fish settlement and conservation and might, therefore, serve as a management tool to aid ecosystem recovery.

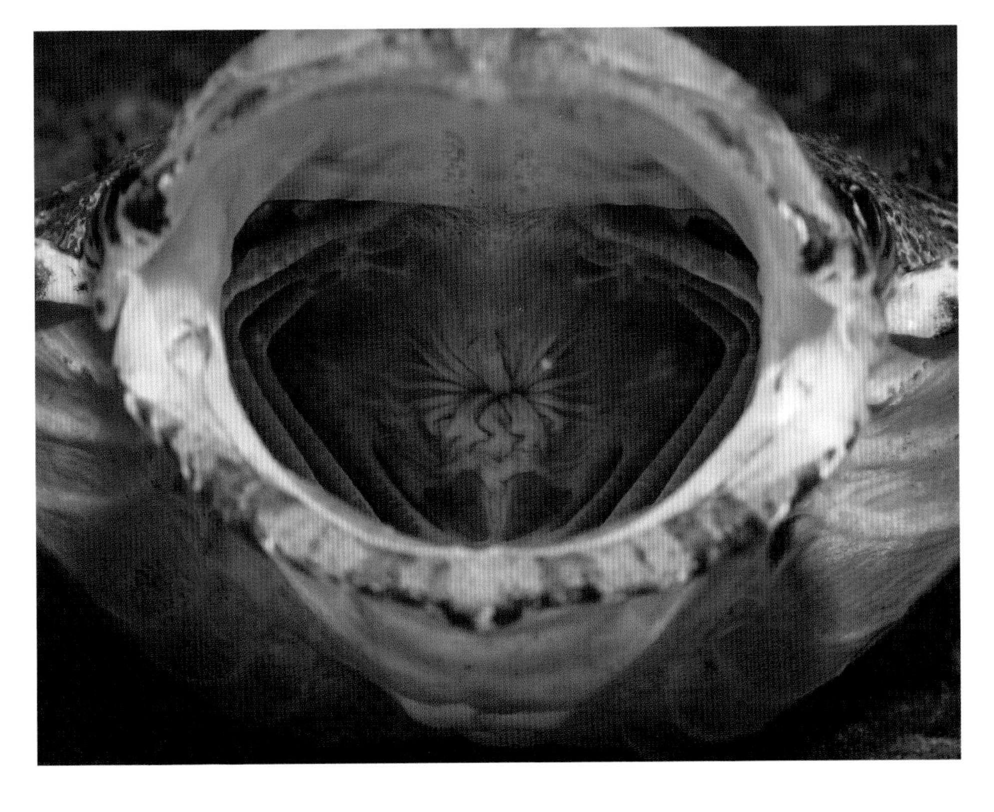

< Yawning allows additional water to flow into fishes, cleaning and oxygenating their gill filaments, and allowing for better breathing while stationary. Crocodile fish. Eilat, Israel.

37 https://www.bbc.com/news/av/newsbeat-43968475/this-is-what-coral-reefs-sound-like (retrieved March 6, 2021).

38 https://www.youtube.com/watch?v=XC6I8iPiHT8 (retrieved March 6, 2021).

39 The benthic zone is the ecological region at the lowest level of a body of water.

> Green sea turtle among big sponges on a reef wall. A new hypothesis suggests that mutualistic symbiotic magnetotactic bacteria might be the underlying mechanism of magnetic field-based navigation of many animals, including sea turtles. Sensing a magnetic field as weak as the Earth's is extremely challenging and the sensory mechanism behind it is widely debated and enigmatic. Bunaken, Indonesia.

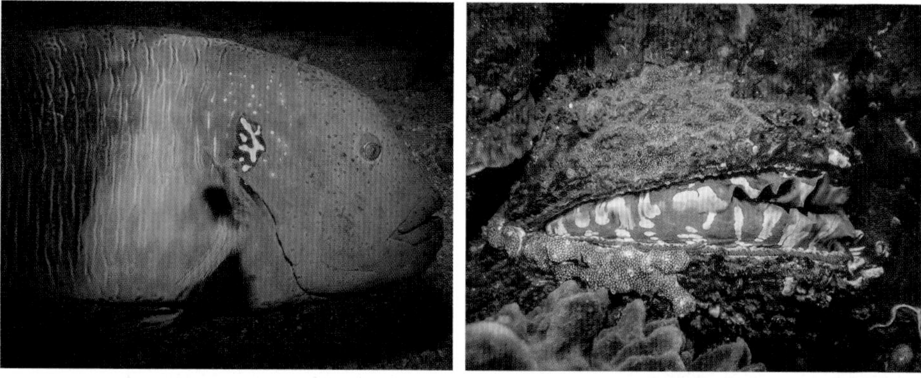

>> The spiny shell of a variable thorny oyster. Puerto Galera, Philippines.

> Broomtail wrasse. Eilat, Israel.

> A school of Klein's (blacklip) butterflyfish mobbing fish eggs over a diverse coral reef. Verde Island, Philippines.

>> Blue ribbon eel. Dauin, Philippines.

< School of anthias.
Verde Island, Philippines.

^ Whitetip reef shark. In some cases, the presence of sharks and other large predators may be an indication of a reef's overall good health. Bunaken, Indonesia

Coral Colonies: Do They Constitute Superorganisms?

Siladen is a tiny island located in Bunaken Marine Reserve, not far from the northwestern shore of Sulawesi, Indonesia, near the center of the Coral Triangle. It is an enchanting place with beautiful white beaches, a few minutes away from one of the world's most attractive dive sites, one which includes almost 400 coral species, 2,000 fish species, sharks, turtles, and many other life forms.

Diving amongst the beautiful coral reef walls in the Bunaken Marine Reserve is a memorable experience. The serene atmosphere of the island induces calmness and a philosophical state of mind, allowing me to ruminate over fundamental questions of life while thinking about the reef. While seeing one of the most impressive "walls of tiny mouths" in the world, fundamental questions echoed in my mind. How should one refer to a massive brain or table coral, one which measures several meters in size? Is it a single organism? A massive aggregate of polyps? Something in between? These questions are not just philosophical questions; they are at the core of evolution theory, relating to topics such as the boundaries of an organism, cooperation versus competition, and the basic unit upon which natural selection acts.

φ φ φ

The concept that single-species groups, including human societies, and multi-species communities, can possess the characteristics of a single organism, sometimes known as a superorganism, has been considered, along with the history of ideas, since ancient times. In Hindu cosmology, the four major castes are described as having descended from different body parts of a primal giant created by the gods. These castes are expected to work in collaboration, much like the organs of a body. A similar idea can be found in Plato's *Republic*, where Socrates makes an analogy between the city-state and man. The metaphor of superorganism was also part of the discourse of the dominant political ideologies of the twentieth century, including Communism, Nazism and Liberalism. For example, during World War II, many American scientists associated group selection with fascist ideologies, while individual selection was associated with democracy. Likewise, during the Cold War, cooperation for the benefit of the whole was linked with group conformity and communism, while diversity and individual freedom were associated with democracy.

Prior to any discussion about whether coral colonies constitute superorganisms, it is essential to define the concept of holobiont, and distinguish it from the term superorganism. Although the terms are sometimes used interchangeably, this distinction is important.

In a nutshell, and for the sake of our discussion, a holobiont is composed of different species that live in symbiosis and create single ecological units. A superorganism, by contrast, is composed of many individuals of the same species that act as a single organism. In both holobionts and superorganisms, multiple members work for the benefit of the collective unit.

Biological hierarchy and holobionts

One theory about how complex life came to be on Earth is the endosymbiotic theory articulated by Lynn Margulis (1938–2011). This theory proposes a hypothesis regarding the emergence of eukaryotic cells, the primary building blocks of all multicellular organisms.[40] According to this theory, the symbiotic partnership between two separate primitive cells enabled them to thrive, develop organelles for specialized tasks, and

40 Eukaryotes are organisms whose cells have a nucleus surrounded by a membrane. Prokaryotes, which include bacteria and archaea, have no membrane around their nucleus, nor around other organelles.

eventually give rise to complex multicellular organs and bodies.[41] Margulis coined the term holobionts to describe ecological units created by different species. She suggested that physical association between individuals of distinct species for substantial portions of their lives constitutes symbiosis. A holobiont inhabits an ecological niche, adapts, and may establish the organizational level (unit of natural selection) that is subject to natural selection forces. An excellent example of a holobiont is the human body; it consists of bacteria, viruses and sometimes larger organisms such as worms and fungi. The number of symbiotic microbes in the human body is in the same order of magnitude as the number of cells. The genome of these bacteria is a few orders of magnitude higher compared to the human genome, which means that most of the genome in our body is not "ours." Many of these bacteria influence various aspects of our behavior, preferences, and, needless to say, physiology. In the words of Margulis, "We are all of us walking communities of bacteria. The world shimmers, a pointillist landscape made of tiny living beings."[42]

Biological hierarchy
From atoms to Gaia

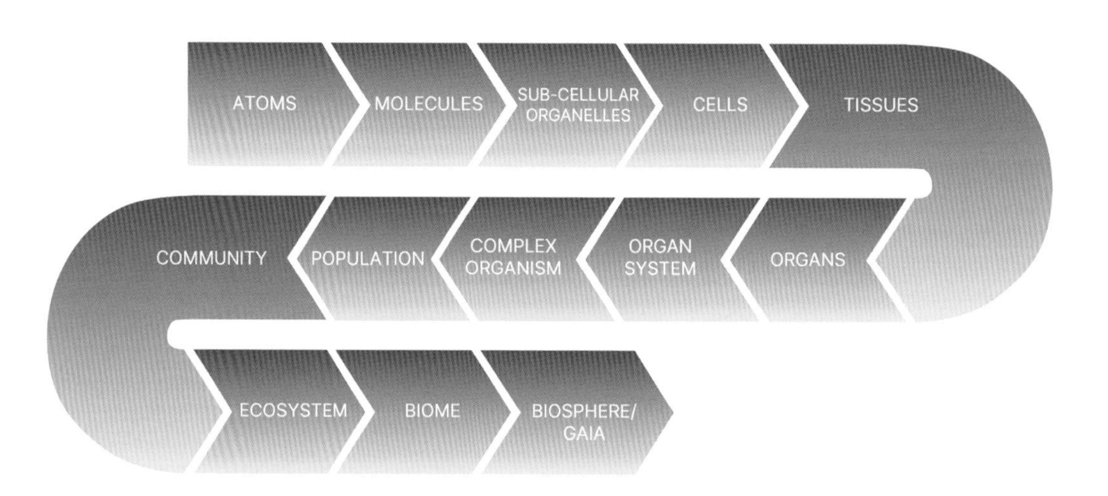

41 The endosymbiosis theory suggests that the evolution of cells is the result of primordial synergy, in which organelles such as mitochondria evolved from a cooperation between bacteria and algae that eventually became incorporated into an individual cell. Mitochondria, the "powerhouse" of most nucleated cells, are responsible for cell respiration; they include separate genetic material and reproduce independently, and at different times, from the rest of the cell. Margulis suggested that originally, mitochondria were independent cells that were incorporated into other organisms and became an integral part of them.

42 From Margulis & Sagan's 1997 book, *Microcosmos: Four Billion Years of Microbial Evolution*.

When challenged by environmental perturbations such as natural selection, a holobiont can adopt strategies and increase fitness in a way that is not possible for any species by itself. Natural selection will then favor the holobiont form, which confers increased fitness under the new conditions. This principle demonstrates the importance of symbiotic partnerships as a significant source of new evolutionary trajectories and trait novelty. Some scientists, including evolutionary biologist Ernst Mayr (1904–2005), consider the creation of eukaryotic cells to have been the most important event in the evolutionary history of the organic world. Similarly, but at a higher level, futurist James Lovelock (1919–2022) described Earth's entire biosphere as a kind of self-organized holobiont that self-regulates its atmosphere to be conducive to life. Lovelock called this the Gaia hypothesis, suggesting that Earth is "the largest living organism in the Solar System."[43]

Biological hierarchy, therefore, can be illustrated as a continuum, from the atom level to the Gaia, in the following manner: In hierarchical systems such as those described above, each level of the biological structure adds unique emergent properties that were not present in its previous parts, emerging as a result of an interaction between its components.[44] It is tempting to add to this biological hierarchy the metaphysical layer proposed by Jesuit priest Pierre Teilhard de Chardin (1881–1955). De Chardin coined the term Omega Point as the final point of evolution, the point where humans' cognitive layer and consciousness, what he called the Noosphere, progresses towards unification with God and independence from a physical being. However, de Chardin's metaphysical theory relates to cooperation between members of a single species and not between different species, as proposed by the concept of the holobiont. In other words, collaboration between members of the same species can be considered a special case of a holobiont: a superorganism.

The coral colony as a holobiont

A closer look at reef-building corals reveals that they incorporate thorough, permanent interactions between various life forms; in other words, corals are holobionts. Besides the algae that reside within them, corals are crowded with microscopic life forms: bacteria, viruses and a diverse collection of single-celled organisms called protists, such as the amoeba. It has been found that sediments around coral reefs contain

43 From Lovelock's 1986 article, "Gaia: the world as living organism."

44 This principle is called holism, meaning that a higher level of order cannot be explained by examining its component parts separately.

approximately 10,000 times more bacteria than the seawater that surrounds them. Moreover, most of these various life forms live in a symbiotic web of interactions that supports the functioning and resilience of the entire coral reef habitat. It is possible that the bacteria deliver nutrients to corals, similarly to the way bacteria in the mammalian gut facilitate the absorption of nutrients during digestion. These bacteria may also be involved in coral metabolism as well as releasing antibiotics that help the corals cope with certain diseases.

The potential of thorough future integration between the symbiotic components of the holobiont exists because of the interdependencies, high internal cycling and structural complexity. This potential integration might also accelerate due to the increase in external human-caused selection pressures.

Superorganisms in biology

The term superorganism in its biological context was introduced, in the first half of the last century, by entomologist William Morton Wheeler (1865–1937), who used it to describe a colony of eusocial insects, such as ants and termites.[45] He found that the social behavior of these animals was remarkably complex and that they acted as a single organism, a so-called superorganism.

The "super" in superorganism reflects a higher organizational level in a group of the same species. Members of the group are responsible for performing specific tasks (such as reproduction, feeding, movement and defense), much like organelles in cells or organs in the body.

In their article "Reviving the superorganism," philosopher Elliot Sober (1948–) and evolutionary biologist David Sloan Wilson (1949–) proposed a more formal definition of a superorganism: "A collection of single creatures that together possess the functional organization implicit in the formal definition of organism. Just as genes and organs do not qualify as organisms, the single creatures that make up a superorganism also may not qualify as organisms in the formal sense of the word."

The term superorganism provokes curiosity because it embodies a paradox based on the absence of a clear boundary between the individual organism and the superorganism. It resembles, in a way, Zeno's arrow paradox, where an arrow in flight can never reach its target because at every instant, there is no motion occurring,

45 Eusociality is a high level of social relationships among animals. It is characterized by, among others, cooperative brood care of juveniles and a reproductive division of labor (into reproductive and non-reproductive castes) for the benefit of the entire group.

and it occupies a fixed position.[46] In the same manner, we are unable to identify the specific point at which an organism should be regarded as a superorganism.

< Termite nest. Termites are eusocial insects and considered superorganisms. They express complex social behaviors such as living in groups, cooperative care of juveniles, reproductive division of labor, and more.
In some cases, individuals of at least one labor caste lose the ability to perform tasks that are performed by another labor caste.
Omo Valley, Ethiopia.

As suggested by Wheeler, the best examples of superorganisms can be found in eusocial insects, which account for 15–20 percent of Earth's animal biomass, exceeding the biomass of all vertebrates. Eusociality is mainly characterized by cooperative brood care, a high dependency between group members, and a division of labor between reproductive and sterile individuals.[47]

In this respect, the human body can be considered a superorganism, where our organs and cells are so integrated that we barely notice signs of coloniality in our body operation. A different way to view our body is as a collaborative group of highly differentiated, genetically identical, but physiologically mostly integrated cells, all working for the benefit of the collective: the human body.

---

46 Zeno's arrow paradox argues that motion is impossible. Zeno concluded that since time is composed of moments, and an arrow travels no distance at any friction moment, the arrow never moves.

47 In marine taxa, eusocial species are far less common than in terrestrial systems. Probably the only known eusocial animals in the sea are some species of snapping shrimp (genus *Synalpheus*), that live in obligate associations mostly with sponges and crinoids and use the host structure as a nest. In each group of snapping shrimp, there is no more than one reproductive female, similar to a typical termite colony that has a single pair of reproductives: the king and the queen.

Until recently, the prevailing paradigm in modern biology has rejected the concept of superorganisms, adhering to a reductionist approach that explains all adaptations at the individual or gene level, known as the "selfish gene" concept.[48] Many evolutionists support the idea known as individual selection, which holds that the individual is the embodiment of selection, and communities of organisms as well as human societies are mere collections of organisms, without themselves having the features of an organism. This approach contradicts what we know about the emergent properties of systems that cannot be explained by the sum of their parts.[49] In opposition to this, scientists who support the concept of superorganisms argue that superorganisms are more than just a theoretical theme and can literally be found in nature; for example, in the eusocial insects mentioned above.

In this regard, colonies of different cnidarian polyps can provide us with an interesting angle, since they represent a variety of life forms that are positioned along the continuum of the individual organism and the superorganism; from a single polyp to a colony of polyps. Thus, the analysis of members of the phylum Cnidaria might help us answer key questions about how superorganisms arise from the combined operation of tiny organisms.

Coral colonies as superorganisms

Evolutionary biologist Stephen Gould (1941–2002), who was also interested in the question of whether coral colonies constitute a superorganism, suggests seeing superorganisms as a continuum rather than a discrete object with clear boundaries. In his book *The Flamingo's Smile: Reflections in Natural History*, he wrote:

> *Nature, in some respects, comes to us as continua, not as discrete objects with clear boundaries. One of nature's many continua extends from colonies at one end to organisms at the other. Even the basic terms – organism and colony – have no precise and unambiguous definitions. We may, however, use the two criteria of our vernacular as a guide. We tend to call a biological object an*

48 In *The Selfish Gene*, author Richard Dawkins defined the gene as a "selfish" (independent unit) replicator that uses the organism as an instrument for its own survival and propagation. A well-known phrase used by Richard Dawkins and attributed to British author Samuel Butler (1835–1902) that describes this idea is: "The chicken is the egg's way of making another egg."

49 Emergent properties in coral reefs arise through the integration of processes and interactions between reef dwellers. In other words, some of the reef system's properties cannot be explained by its parts, and emerge only at the reef's system level.

organism if it maintains no permanent physical connection with others and if its parts are so well integrated that they work only in coordination and for the proper function of the whole. Most creatures lie near one or the other end of this continuum, and we have no trouble defining them as organisms or colonies.

Two significant pieces of evidence support the hypothesis that coral colonies operate as superorganisms, or at least are very close to it on the individual–superorganism continuum. The first is the characteristics of the colony and the interconnection between the polyps that compose the colony; the second relates to the labor division (polymorphism),[50] and other characteristics of superorganisms that are evident in some cnidarians, including a few species of coral that might hint at potential future evolutionary directions of coral colonies.

Most reef-building coral species form colonies via asexual reproduction (budding) of polyps and are therefore genetically identical. Budded polyps usually originate from internal division of existing polyps, or through the budding of polyps from tissues adjacent to existing polyps. In most colonies, these polyps remain interconnected and integrated via nerve and muscular networks, within the thin layer of tissue that overlays the colony's skeleton. Translocation of organic material between polyps can occur in two ways: diffusion via a concentration gradient or through active transport via the gastrovascular system.[51] This interconnected system functions in a variety of tasks, including a hydrostatic skeleton that allows the colony to grow; a transport system for the distribution of materials and gases between polyps; an excretion system for waste metabolites; and a system that controls the morphological development of colonies. The bottom line is that polyps within a colony share resources for the benefit of the entire colony, although this characteristic alone is insufficient to constitute a superorganism.

The second supporting evidence is based on the existence of some cnidarian species that demonstrate a high degree of specialization and functionality between genetically identical units. This contrasts with the basic coral unit, the polyp, where each polyp is responsible for performing all tasks (reproduction, preying, digestion, etc.) and may suggest a possible evolutionary direction of coral polyp colonies.

---

50 In the context of cnidarians, polymorphism could be the existence of individuals (zooids) of a single species, in more than one physiological and morphological form and in different functions, that act as a well-organized division of labor.

51 In many coral species, the gastrovascular system is composed of two interconnected networks of canals filled with fluid and circulating cells.

In colonial hydrozoans (relatives of corals that belong to the Hydrozoa class), individual zooids (the equivalent of coral polyps) form specialized structures that are part of a bigger colony. In some cases, there are five types of functional zooids in the same colony, each type specialized for a specific action, such as feeding, reproducing, protecting and sensing.

> The Portuguese man o' war (*Physalia physalis*) is a community of cnidarian organisms that live at the sea–air interface. It consists of functionally specialized zooids and compound structures that are homologous to free-living polyps. Each assemblage of zooids works together in capturing prey, feeding, locomotion, floating and reproduction, functioning as a single animal. Historically, the Portuguese man o' war started as a colony of discrete organisms, each capable of performing a full set of functions. Over time, the colony became integrated, with differentiated units that are subordinate to the whole and act as a single individual, also known as a superorganism.

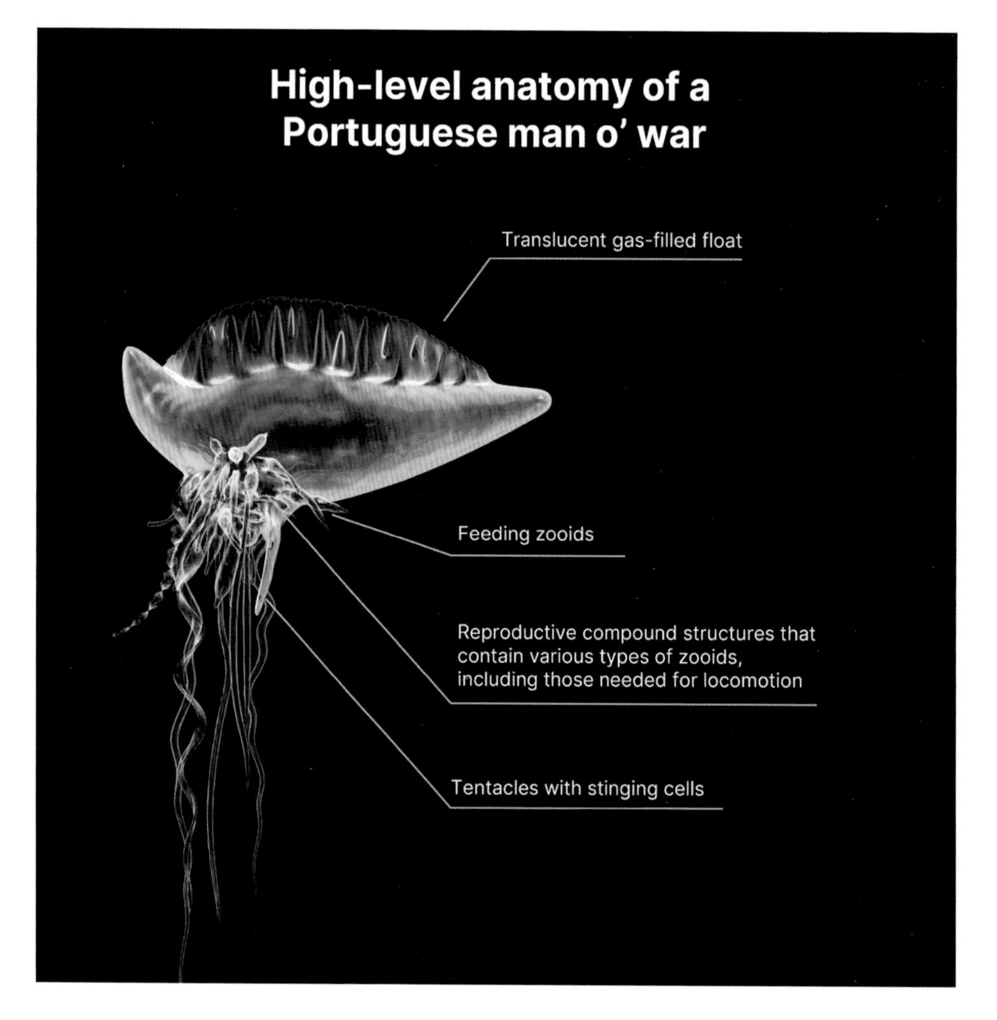

Division of labor can be found in fire corals, which, despite their great similarity to stony corals from the Anthozoa class, are actually hydrozoans, and also in the Portuguese man o' war, where each colony is composed of several zooid types that are integrated to act as one individual. The zooids of the Portuguese man o' war are no longer individuals and cannot survive as separate creatures since they are specialized in a single task, similar to organs in our body. Similarly, there are non-reef-building cnidarians, such as sea pens, that exhibit polymorphism and include up to five different morphological polyps that perform different functions. In other words, the

< Sea pens are octocorals that exhibit polymorphism. That is, they have multiple types of polyps (zooids), that perform different functions. Dumaguete, Philippines.

< Fire coral (*Millepora*) are not true corals but are part of the Hydrozoa class, Sinai Peninsula, Egypt.

phylum Cnidaria is an excellent example of the continuum between solitary living and integrated colonies of genetically identical polyps/zooids that perform various tasks. While some cnidarian species demonstrate some degree of differentiation and specialization, for most coral species a polyp is still a completely functional, independent creature that has not developed any specialty at the expense of its independence as an individual.

Lynn Margulis proposed that multicellular organisms evolved from the symbiosis of distinct species of primordial single-cell organisms that eventually led to the incorporation of their genomes into one multicellular organism. In addition, multicellular organisms have evolved as a result of the differentiation of genetically identical cells to function-specific cells. In the same manner, coral colonies may become superorganisms through two unrelated avenues: through the physiological integration of the biont components (different species) and/or by labor division of genetically identical coral polyps (same species).

All of the above might indicate that some species of coral colonies may be on their way to becoming superorganisms in the far future; but at present, reef-building corals are better described as holobionts.

> A Bennett's (yellow) crinoid (feather star) spreads its arms to catch plankton. Puerto Galera, Philippines.

Resolving Darwin's Paradox

In April 1836, after four years of voyaging, Charles Darwin, a passenger on HMS *Beagle*, was astonished by the crowded pageantry of the atoll reefs of the Cocos (Keeling) Islands, southeast of Java, Indonesia. This beautiful pastoral island inspired Darwin to ask a fundamental question about coral reefs. What struck Darwin was not just the beauty of the reef, but also the "infinite number" of organisms dwelling in it. On land, the diversity of flora and fauna on the Cocos Islands was not impressive at all. Yet in a non-continuous manner, just a few meters away from the land, in the coral reef surrounded by low-nutrient ocean water, an amazing diversity of life forms thrived. What are the reasons behind the high productivity of the coral reef in low-nutrient ocean water? This mystery, now known as Darwin's Paradox, was posed by Darwin back in 1842 and has yet to be solved.

φ φ φ

Habitat productivity is defined as the biomass produced per square meter per year.[52] The productivity of the coral reef habitat is 20 times higher than that of the open ocean and is also remarkable compared to terrestrial habitats. In some cases (mainly in atolls and islands), coral reefs are surrounded by oceanic marine habitat equivalent to a desert, and as a result, a reef is often called an "oasis in the desert." This paradox can only be explained by processes ensuring efficient capture, retention, and recycling of energy and nutrients.

Darwin's Paradox is a derivative of one of the most fundamental questions science has ever coped with: What is life? In other words, what is the difference between an animal cell and inanimate objects? At first glance, this question may seem unrelated to Darwin's Paradox, but the same fundamental physical law that distinguishes between animate and inanimate entities also stands behind one of the essential explanations for coral reefs' high productivity in low-nutrient environments.

Nobel Prize-winning physicist Erwin Schrödinger (1887–1961) tried to answer this question in a collection of lectures delivered in 1943 under the title "What is life?" Schrödinger rejected the idea that life is a mystical thing characterized by a vital impulse (*élan vital*). On the contrary, he argued that life is a process that can be described by mere physical and chemical laws that govern the interaction between energy and material.

To better understand Schrödinger's arguments, we need to go back to the first and second laws of thermodynamics, and focus on life as a process of energy transformation. The first law of thermodynamics is fairly simple. The amount of energy in the universe is constant and eternal. Energy is conserved and neither created nor destroyed. For example, the amount of energy in the first instance of the Big Bang 13.8 billion years ago is precisely the amount of energy that exists today. In a way, the history and evolution of the universe might be described as the history of energy transformation.

52 Biomass is the mass of living organisms in a defined area at a given time. It includes microorganisms, plants and animals.

The second law of thermodynamics is more elusive but no less significant. It states that the natural tendency of any isolated system (such as the entire universe) is to degenerate into a more disordered state. In other words, the disorder (entropy) of an isolated system always stays the same or increases.

This principle explains why life is unique. Living creatures, some of them, are very complex, avoid decay, and actually work in the opposite direction; that is, they increase internal order and complexity, thereby lowering entropy. The process of avoiding decay that characterizes living creatures contrasts with the continuous decay process of inanimate objects. Decay avoidance is what makes organisms so enigmatic and was the pillar for the longstanding metaphysical belief that there is a supernatural force behind every organism. The generation of order and complexity is done by using external energy such as sunlight in phototrophs,[53] or food in herbivores and carnivores. On a universal scale, the second law of thermodynamics is not violated, and the amount of disorder (entropy) in the universe always increases with time. That is to say, living things are enclaves of order; by using energy, they "import" order (negative entropy) and "export" disorder.

A coral reef is no exception. At first glance, this "wall of tiny mouths," inhabited by a variety of marine creatures, seems to violate the second law of thermodynamics. That is to say, there are more consumers than producers in the reef, and it seems that there would be insufficient energy to support the high-level order and complexity. Unlike terrestrial habitats that contain a large number of plants (producers) and relatively low number of herbivores and carnivores (consumers), the number of plants in coral reefs seems to be small, while the number of herbivores and carnivores is vast. So how do these consumers feed?

A more in-depth look resolves the mystery. The corals' symbiotic algae, which live within the coral, convert a massive amount of sunlight energy into food that is later consumed by the coral polyps. In addition, the coral also feeds on external prey captured by the polyps. Coral colonies are actually composed of internal producers (algae) and consumers (polyps) that enable them to efficiently recycle the limited supply of external nutrients.

In sum, coral colonies and other animals that possess photosynthetic algae demonstrate high energy efficiency. The internal symbiosis between producers and consumers enables them to maintain relatively high productivity in low-nutrient

53 Phototrophs are organisms that can use visible light (photon capture) as a primary source of energy for metabolism in a process known as photosynthesis. Chemotrophs, by contrast, are organisms that produce energy from the oxidation of organic compounds.

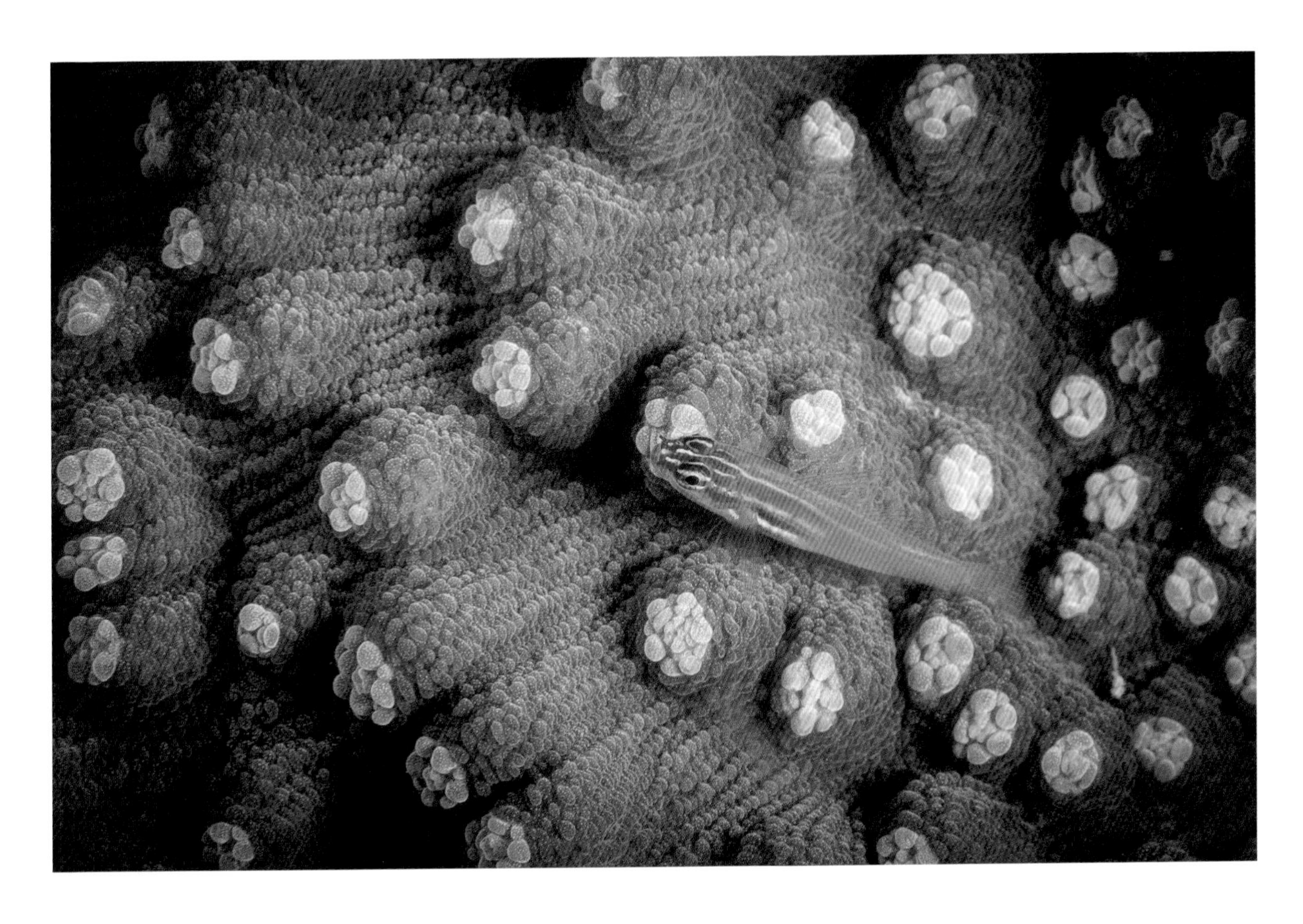

^ A blackbelly dwarf goby (*Eviota atriventris*). Lembeh Strait, Indonesia.

ocean water. This explanation, however, does not entirely resolve Darwin's Paradox since additional vital nutrients (mostly nitrogen and phosphorous) are required to sustain the reef.

In addition to the above hypothesis, scientists have suggested other solutions to address the mystery of Darwin's Paradox. For example, they showed that, in addition to the transfer of dissolved organic matter via bacteria to the reef's fauna, sponges make this organic material available to fauna by rapidly expelling filter cells as dead particulate organic material trash (detritus). This material is subsequently consumed by the reef's fauna.

Further research has shed light on the contribution of an effect called the Island Mass Effect. According to this effect, hotspots of rich phytoplankton,[54] an essential source of energy in the marine environment, are the result of positive feedback that enhances phytoplankton proliferation within the island's vicinity. There are multiple factors behind this positive feedback, including island morphology, water depth, and

54 Self-feeding components of the collection of organisms that live in a water column.

> Small *Mexichromis* (*Durvilledoris*) nudibranch. Lembeh Strait, Indonesia.

⌄ West Wind *Hypselodoris* nudibranch. Lembeh Strait, Indonesia.

local human impact. For instance, reefs continually create gradual slopes that enable the movement of nutrients from the deep sea to the shallow waters.

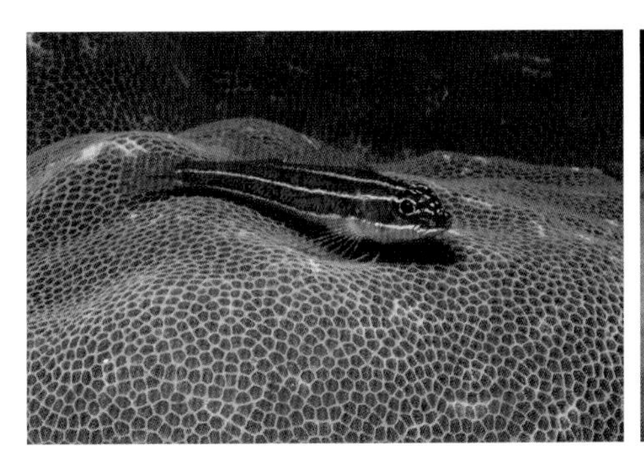

« Tropical striped triplefin blenny on coral. Verde Island, Philippines.

‹ Schultz's pipefish (gilded pipefish). Eilat, Israel.

Another complementary explanation for Darwin's Paradox comes from the existence of a massive number of small cryptobenthic reef fishes such as gobies and blennies.[55] In some localities, these fishes may account for more than 40 percent of the reef's biomass. Often overlooked, these fishes are characterized by their short life cycle, and the production of vast amounts of larvae that remain rather close to the reef and serve as a significant part of the food web, estimated to be almost 60 percent of the consumed reef fish biomass.

It seems that Darwin's keen observations, which led to a longstanding scientific debate, are now being explained by multiple complementary theories that highlight the uniqueness of the coral reef habitat and provide us with a broad explanation of its magnificent, yet puzzling, existence.

55 Cryptobenthic fishes are fishes that hide or are camouflaged, and live near or on the floor of the sea.

"At the antipodes of every mind lay the Other World of praeternatural light and praeternatural colour, of ideal gems and visionary gold. But before every pair of eyes was only the dark squalor of the family hovel, the dust or mud of the village street, the dirty whites, the duns and goose-turd greens of ragged clothing. Hence a passionate, an almost desperate, thirst for bright, pure colours; and hence the overpowering effect produced by such colours whenever, in church or at court, they were displayed."

Aldous Huxley, *Heaven and Hell*, 1956

CHAPTER 3

The Reef Kaleidoscopic View
Why Are Coral Reefs So Colorful?

« A garden of ascidians (sea squirts). Ascidians, often confused with sponges, are complex filter feeders that can pump tens of liters of water through their bodies each day. Lembeh Strait, Indonesia.

Anyone who has snorkeled or dived in a coral reef remembers their first encounter with the brilliantly colored fishes that inhabit this magnificent habitat. Compared to other habitats on the planet, coral reefs have an exceptional place, and are imprinted in our minds as a symbol of beauty and natural richness. During my many years of diving, I have often tried to find the root causes behind the colorful nature of coral reefs.

British essayist and philosopher Aldous Huxley (1894–1963) dedicated many years of his life to exploring the boundaries of human consciousness. As part of his quest, he experienced the hallucinatory material mescaline, which is found in the peyote cactus, a plant used for thousands of years by the indigenous peoples of North America. Native Mexican shamans used it in visionary rituals that open a pathway to other deities and realms.[56]

In his philosophical essay *Heaven and Hell*, Huxley described and analyzed the antipodes of human consciousness, regions outside the system of mundane conceptual thought. Accessing these areas can be achieved by activities such as meditation, fasting, self-flagellation, and also by using chemicals such as mescaline to create an altered state of mind. Huxley asserted that our attraction to colorful forms and views results from our desire to visit and enjoy brilliantly colorful, unmapped unconscious visionary experiences. In fact, throughout human history, attraction to bright colors has been considered the essence of artistic beauty. For many generations, it was believed that under certain circumstances, colorful objects made of shiny, pure colors could alter one's state of mind in the direction of its unconscious antipodes.

56 A classic book describing psychedelic experience with peyote is *The Teachings of Don Juan: A Yaqui Way of Knowledge* by Carlos Castaneda (1968).

> Colorful Buddhist mask dance in Bhutan. Mask dances often represent legendary mystical stories with spiritual significance. They are based on episodes from the life of Guru Rinpoche, who introduced Buddhism to Bhutan and is considered the patron saint of Bhutan and neighboring Buddhist cultures. Huxley argues that our natural attraction to colorful scenes is the result of our desire to "visit" and enjoy those brilliantly colored, unconscious antipode areas. Thimphu, Bhutan.

Huxley's ideas represent mystical and philosophical viewpoints that hint at the reasons behind our attraction to colorful landscapes. However, human perception is clearly not the only answer to the question of why so many reef creatures are vivid and boldly ornamented. The reef does not exist for mere human pleasure, so there must be other reasons for its abundance of color. To grasp this complicated topic, we need to add some conceptual layers.

The first layer sets out high-level evolutionary perspectives and constraints, which can be summed up as communication or signaling (reproduction or warning coloration, for example), and anti-communication (camouflage). The second layer describes the unique properties of light and the color spectrum in an underwater environment, which is different from light on the land. The third layer relates to the guiding principles of camouflage; some of which are counterintuitive. The last layer comes from cognitive psychology and is, to a certain extent, complementary to Huxley's anthropogenic ideas. This layer provides insight regarding our tendency to portray these habitats in a colorful manner, which may well be different to how they are perceived by reef animals. All of these layers provide a comprehensive explanatory framework that addresses this chapter's central question, "Why are reef creatures perceived to be so colorful?"

The Constraints of Natural Selection

Ever since discussions of evolution theory began in the nineteenth century, they have included reflections on the issue of animal coloration. Alfred Russel Wallace was among the prominent scientists who suggested a hypothetical answer to this question. In his 1879 article "The Protective Colours of Animals," Wallace proposed that the bright colors of reef fishes may help them merge into the equally colorful environment. "Brilliantly-coloured fishes from warm seas are many of them well concealed when surrounded by the brilliant sea-weeds, corals, sea-anemones, and other marine animal[s]."

Another suggestion regarding this puzzle came from Austrian ethologist Konrad Lorenz (1903–1989), who proposed that conspicuous colors help fishes identify their species. Lorenz asserted that the highly diverse and multi-niche reef environment is essential when considering breeding and interspecies territorial competition.

Broadly speaking, the survival of any animal species requires three elements that may be strongly influenced by color: feeding (camouflage serves as part of the preying mechanism), avoidance of predators (camouflage and warning coloration), and reproduction (signaling). For many sea animals, the need to disappear sometimes forced by the first two of these requirements conflicts with signaling.

It seems that both Wallace and Lorenz were right, but only to a certain extent. As I elaborate later, additional factors influence animal coloration, which is mainly derived from the properties of light in water and the guiding principles of camouflage. These factors enable reef animals to bridge the conflicting demands of appearance and camouflage. In order to address these evolutionary constraints, reef creatures can be viewed along an "appearance" vector that includes completely cryptic animals such as flatfish or frogfish, at one end, and conspicuous fishes such as lionfish, at the other end. Between these two endpoints, there are various ways in which the adaption of colors and patterns enables fishes to be alternately camouflaged and conspicuous, at various times, and in various places and scenarios.

One more fact that needs to be considered regarding the predation risk of boldly ornamented animals is reef morphology. Coral reefs provide a structured habitat with a vast surface area and many niches for various species. It is a complex, tangled habitat that provides both shelter and nursery grounds, so colorful and conspicuous animals can easily find refuge, thereby significantly reducing their predation risk.

Communication and signaling

Coloration is one of the most common ways of communication and anti-communication in the animal kingdom. Wallace originally contemplated the concept now known as aposematism to describe prey that combines warning displays with secondary defenses. Aposematism describes the use of highly conspicuous colors to communicate distastefulness, toxicity and warning to potential predators; in other words, "If you attack me, you are going to regret it later." The design of aposematic signals evolves to increase the detectability of an unpalatable species. This warning coloration enhances the predator avoidance learning curve and decreases recognition error by predators. Some research has shown a positive correlation between color pattern brightness and the level of toxicity. Boxfishes, blue-ringed octopuses, and nudibranchs are all well known for such warning displays. In contrast to the hidden behavior of camouflaged animals, animals with aposematic displays are generally characterized by confident, indiscreet behavior.

> Lionfish. Eilat, Israel.

Another interesting piece of the puzzle that might help explain why reef fishes are so colorful is that coloration and patterns may depend on the presence of other species. A good example can be found in two closely related species of butterflyfish, the reticulated butterflyfish and Meyer's butterflyfish, that live in close proximity to each other in many reefs, and need different colors to stand out. Speciation between the two took place less than a million years ago. Interestingly, not only did it result in different color patterns (and similar body shapes); a recent study found that in reefs where the ranges of these species overlap, the differences in color patterns were more pronounced than in reefs where the species have less overlap.[57]

Color change is commonly used in reefs to facilitate communication in several forms, including mating, aggressive behavior, and developmental stage. Some species, such as the triggerfish and goatfish, can change color almost as quickly as chameleons. In other fishes, color changes occur at different hours of the day, or during maturation. Color change can also be part of courting, when male or female

‹ The specialized mouth on the poisonous thornback cowfish jets water into the sand, exposing buried small invertebrates. Puerto Galera, Philippines.

fishes become more conspicuous by highlighting existing colors as they try and attract the attention of a potential mating partner. It can also take place during aggressive behavior, as in the case of the blue-ringed octopus.

Reef fishes from several families also show irreversible color change on developmental (ontogenetic) timescales. In many cases, such a change occurs in the transition from the juvenile to the adult stage (common examples are Barramundi grouper and Emperor angelfish) or when a fish changes sex (for example, anthias). Color changes from juvenile to adult are most prominent in territorial fishes. Thus, it may partially function to allow juveniles residing in the same reef territory as

57 "Mystery of color patterns of reef fish solved," ARC Centre of Excellence in Coral Reef Studies, December 5, 2018.

their adults to signal that they are part of the same species, but without encouraging sexual or territorial competition. Color change is also used for anti-communication purposes in frogfishes, octopuses and pygmy seahorses.

> Humpback grouper. In many cases, irreversible color change takes place in the transition from juvenile (left) to adult (right). Note that both juveniles and adults are covered with false eyespots. Lembeh Strait, Indonesia.

A collage of colors: nudibranch warning coloration

Nudibranchs, from the Greek for "naked gills," are a type of soft-bodied fish, often called sea slugs, that belong to the phylum Mollusca. About 3,000 species of nudibranchs can be found in diverse marine habitats, and their reputation precedes them thanks to their spectacular display of flamboyant patterns. The majority of them produce their own color pigments while a few, including the solar-powered nudibranchs, get some of their colors from the algal pigments they ingest.

The two commonest subfamilies of nudibranchs are dorids and aeolids. Dorids look fairly smooth and have bouquet-like gills toward the back of their body that are used for breathing. Aeolid nudibranchs breathe through organs called cerata that spread along their backs. All nudibranchs are hermaphroditic, with both male and female sex organs located on the right side of the body. On the dorsal (back) surface of the head, nudibranchs have chemosensory organs called rhinophores, which are receptors capable of sensing scent, taste and touch.

Unlike most gastropods, more commonly known as snails and slugs, nudibranchs lack the protection of a shell in their adult stage, which means that they are exposed to the threat of predators, but saved from the energy cost of producing a shell (roughly 5 percent of total energy). One would not be wrong to think that nudibranchs' conspicuous colors would attract potential predators; however, some of their fantastic coloration serves as aposematism, signaling to predators that they are unpalatable. About half of all nudibranch species have aposematic coloring; others defend themselves with various types of camouflage.

Interestingly, evolution has produced some aposematic animals, like nudibranchs, that are unable to see their own beautiful colors as they lack the vision features necessary to see colors (they are almost entirely blind). This indicates that their vivid, conspicuous patterns evolved exclusively to address predation risk, and not for any kind of communication or reproduction signaling.

<< Gabriela's tambja. Lembeh Strait, Indonesia.

< Chamberlain's nembrotha dorid nudibranch. Nudibranchs are equipped with chemosensory organs at the back of the head that detect chemicals dissolved in the sea water. These organs, known as rhinophores, are characterized by a large surface area that maximizes chemical detection. Puerto Galera, Philippines.

<< Girdled glossodoris nudibranch. Lembeh Strait, Indonesia.

< Batangas halgerda. Lembeh Strait, Indonesia.

< The simple eyes of this Bullock's hypselodoris dorid nudibranch are only able to discern between light and dark. Puerto Galera, Philippines.

^ Kubaryana's nembrotha.
Puerto Galera, Philippines.

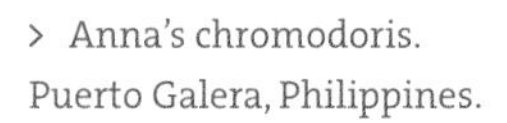

> Anna's chromodoris.
Puerto Galera, Philippines.

<< Bullock's hypselodoris. Lembeh Strait, Indonesia.

< Willan's chromodoris. Puerto Galera, Philippines.

The Properties of Light Underwater

Vision underwater is influenced by the absorption and scattering of wavelength energy by water molecules, as well as plankton, dissolved organic matter, and other particles in the water. The attenuation of colors occurs in horizontal, vertical and diagonal planes, is strongly affected by water conditions, and is different in each water environment.

Because water is such an effective filter of light, the underwater world is viewed differently in just a few meters' depth than it is near the surface. Bright red effectively becomes black when viewed 15 meters below the surface in typical tropical water, while colors visible at shorter wavelengths, such as yellow, green, blue and violet, travel deeper. At greater depths, the predominant visible color is blue.

The reef's most useful color is in the range of near-ultraviolet or violet (350 nanometers [nm]) to yellow (600 nm). Blue-only fishes, such as fusiliers, blend with the blue water background and are conspicuous against the contrasting yellow-brown coral reef background. Similarly, yellow-only fishes such as damselfish are camouflaged on the coral reef background and conspicuous on the blue water background.

Within the edges of the visible color spectrum (ultraviolet to red), yellow and blue are an effective combination and offer sharp contrast, with peaks at different parts of the spectrum and furthest color transmission. They are used for advertising, including warning coloration. Many toxic animals, such as the blue-ringed octopus and some nudibranch species, advertise their toxicity with yellow and blue patterns.

At the same time, and in certain circumstances, the combination of yellow and blue can be camouflage. When blended at a distance, and as the eyes' resolution abilities fail, yellow and blue may combine in an additive color mixture (synthesis) to appear as well-camouflaged gray. Moreover, a yellow–blue combination can also disrupt the fish's body pattern against blue or yellow backgrounds, an effect called disruptive coloration (both effects are discussed further later in this chapter). This gives reef animals the choice to be either conspicuously seen for communication purposes or hidden from predators, depending on the time and place.

Behavior of colors in typical clear water

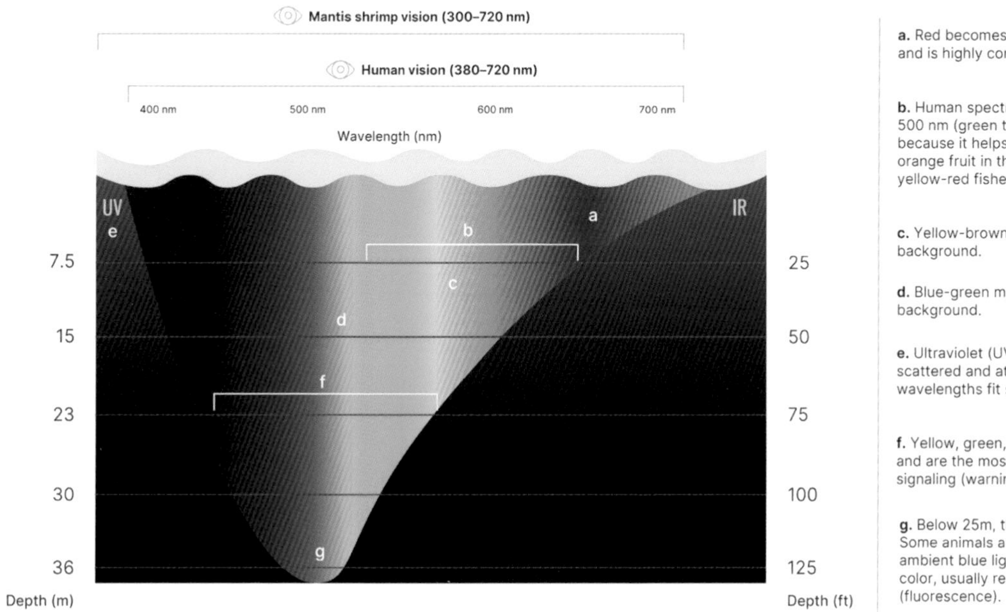

a. Red becomes darker or black in deep water and is highly conspicuous in shallow water.

b. Human spectral discrimination beyond 500 nm (green to red) was probably selected because it helps primates discriminate yellow-orange fruit in the foliage at distance. Thus yellow-red fishes are conspicuous to humans.

c. Yellow-brown matches the coral reef background.

d. Blue-green matches the water column background.

e. Ultraviolet (UV) color (<400 nm) is highly scattered and attenuated by water. These short wavelengths fit short-distance signals.

f. Yellow, green, blue and violet travel deeper and are the most useful colors for long-range signaling (warning coloration).

g. Below 25m, the prominent color is blue. Some animals are capable of absorbing ambient blue light and emitting a different color, usually red, orange or green (fluorescence).

There are also many reef fishes that are red and orange. These colors are more often associated with deeper reefs or with shaded areas since red becomes darker or black and is not seen from a distance. However, while swimming in shallow waters, such fishes are highly conspicuous and might use color for reproductive signaling or warning.

Ultraviolet color is highly scattered and attenuated by water. This means, for example, that visual signals in the ultraviolet that are part of a sexual display could be sent to a nearby mating partner and that these signals would not be visible over longer distances at which predators, should they see them, might ambush. This may explain why so many fishes use ultraviolet or near-ultraviolet colors as sexual communication, and the rich array of fish pigments with reflectance peaks in the UV.

In summary, color contrasts in the water can be achieved by combining short wavelengths (violet to blue) with long wavelengths (yellow to red). Contrast does not necessarily create higher visibility, however, as some color combinations can provide fish with both conspicuousness and camouflage.

The Principle of Camouflage

One cannot help being astonished by the perfect camouflage of some marine creatures, that enables them to resemble other objects in order to hide from potential predators and blend in with their surroundings. In aquatic environments, almost all creatures can be observed from multiple angles, thus camouflage must fit into this environment, and hide the animal when seen from any direction. Camouflage can also serve offensive purposes such as ambushing and attracting prey. It is not surprising that, due to its contribution to animal fitness, camouflage has long been used as a persuasive demonstration of the power of natural selection.

When discussing camouflage, two types of cryptic behavior can be discerned: resemblance and mimicry. As a protective mechanism, resemblance is used by animals that closely resemble a part of their habitat that is of no interest to its enemies, such as the substratum or sedentary animals like sponges or corals. Mimicry is an evolved physical or behavioral resemblance of one organism to another, usually providing a selective advantage to the mimicker. A well-known example of protective resemblance can be seen in flatfishes, which resemble the seabed, and frogfishes, that are well camouflaged near sponges or corals. Frogfishes also show aggressive mimicry by using a lure called an esca, that looks like a shrimp, worm or fish and that attracts prey to within striking distance of the frogfish. A fantastic example of mimicry can be found in the mimic octopus, that can mimic both the appearance and behavior of a vast repertoire of toxic species, including sea snakes, lionfishes, flounders and more.

For animals with slow adaptive camouflage (that is, camouflage that may take hours or even months to display), concealment means finding and moving to the right habitat, with the right lighting and right visual background. Animals can also find the appropriate posture or behavior that conceals them from predators. Alternatively, animals can live with predetermined camouflage patterns that reflect a compromise between the characteristics of several different habitats.

There are two main interrelated but logically distinct archetype mechanisms of camouflage: disruptive coloration, which breaks up an animal's outline with strongly contrasting patterns, and background resemblance, which occurs when an organism's colors are a random sample of the background. A derivative of background resemblance, counter-shading, is described later in this chapter.

The concept of disruptive coloration was first described by American painter and naturalist Abbott Thayer (1849–1921). Thayer's knowledge of painting and familiarity with nature contributed to his articulation of camouflage theory techniques during the First World War. The central idea behind disruptive coloration is the use of large,

> Camouflaged crocodilefish in predatory position. Eilat, Israel.

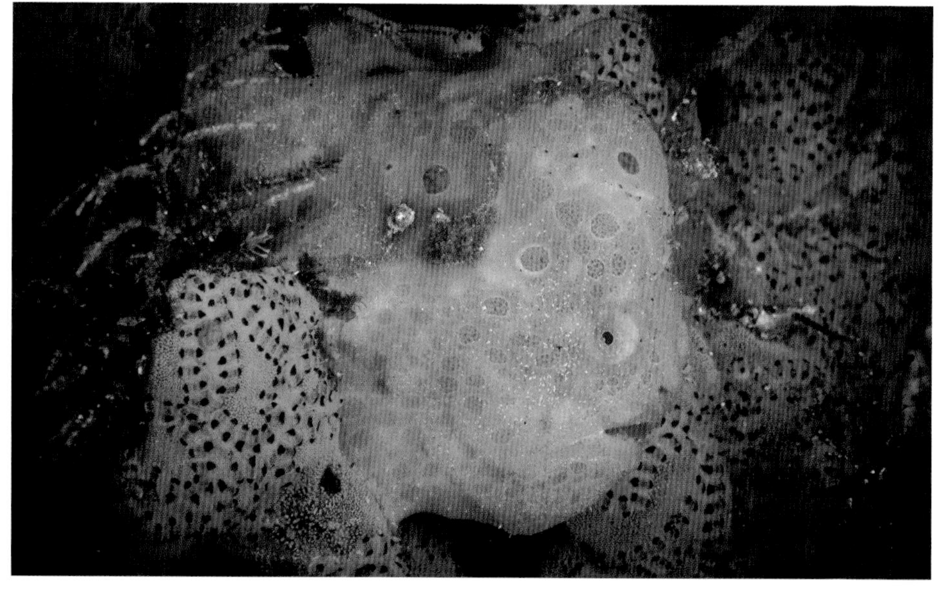

> Painted frogfish, orange phase, adjacent to a sponge. Lembeh Strait, Indonesia.

conspicuous contrasting color patterns that enable an object to merge into its surroundings when viewed against a contrasting background. An excellent example of disruptive coloration is the marine combat uniform made of patterns that break up the outlines of soldiers' bodies. Similarly, using contrasting colors in the form of stripes and dots disrupts the fish body outline, making it almost invisible to predators. In other words, and in a counter-intuitive manner, bold and conspicuous patterns can serve as camouflage. Larger reef fish take advantage of conspicuous spotted or striped patterns for mating, aggressive display and territory ownership. From a distance, those spots and stripes become blurred and camouflaged. Compared with background resemblance and aside from counter-shading, which usually characterizes relatively static benthic animals, disruptive coloration is much more useful for mobile reef fishes.

Another form of animal coloration is additive color mixing. The bodies of parrotfishes, wrasses and some other reef fishes are covered with additive color mixing (for example, blue and red) that contrasts and looks conspicuous at a short distance, but matches the blue-gray background at a distance, concealing the fish from distant predators. Professor Justin Marshall from the Queensland Brain Institute conducted comprehensive research on fish coloration and found remarkable similarities between additive color mixing and pointillism, the revolutionary painting technique developed by French Neo-Impressionists Georges Seurat (1859–1891) and Paul Signac (1863–1935) in the mid 1880s.[58] Seurat and Signac used patterns of distinct dots of pure, unmixed colors to create the effect of different vibrant colors when viewed from a distance. Pointillism was a reaction against Impressionism, which was based on the subjective impression of individual artists. By contrast, pointillism was based on clear methodology and a scientific approach.

˅ The contrasting blue and yellow colors in this emperor angelfish enable communication (at short distance) and camouflage (at long distance), using the principle of disruptive coloration. Eilat, Israel.

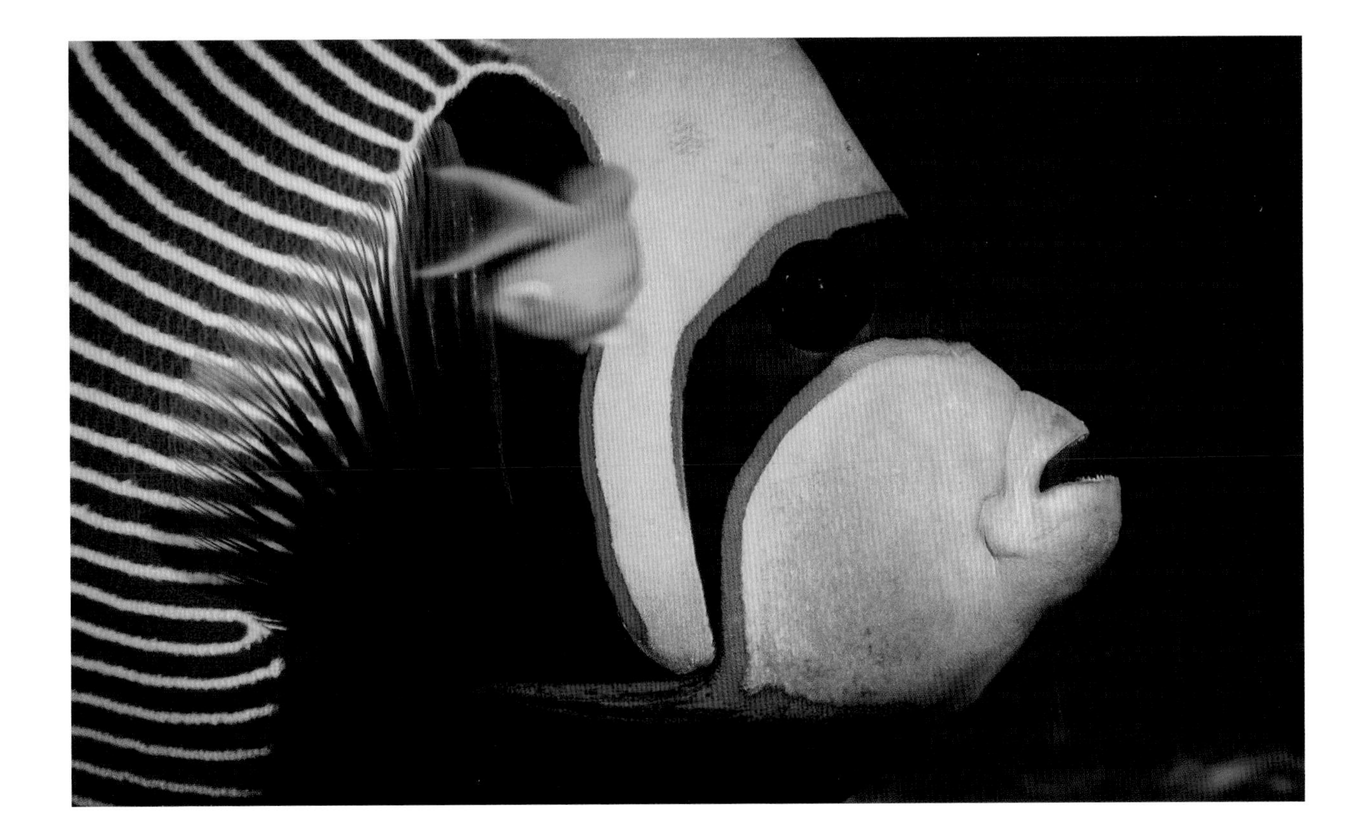

58 Other artists, including Van Gogh, Picasso and Kandinsky, also used this technique.

> *A Sunday Afternoon on the Island of La Grande Jatte* (Seurat, G. [1884–1886]). Pointillism is a painting technique in which small dots of different colors are applied in patterns to form an image. (For full credit information, see Figure credits.)

Important to pointillism, no less than Seurat and Signac, was French chemist Michel Eugène Chevreul (1786–1889). Chevreul discovered that the effect of colors is not solely dependent upon which dyes were used, but how specific different hues were combined. In a nutshell, a painting's visual impact is a matter of optics, not the chemistry of colors. It depends on the harmony and juxtaposition of complementary colors; for example, blue and orange enhance each other's intensity.[59] Today, television and computer monitors use the same principle to represent different image colors using red, green and blue (RGB).

Thayer identified another form of background resemblance, one known as Thayer's law or counter-shading. In Thayer's own words: "Animals are painted by nature, darkest on those parts that tend to be the most lighted by the sky's light, and vice versa."

Counter-shading is widespread in sharks, whose upper parts tend to be dark and lower parts tend to be white, as well as other pelagic or near reef fishes. When the animal is viewed from above, it blends in with the dark blue waters below it, or the dark seabed. When it is seen from below, it blends in with the sunlit waters above it. A similar mechanism, called counter-illumination, can be seen in marine creatures

59 https://www.virtosuart.com/blog/pointillism-art (retrieved March 6, 2021).

such as bobtail squids (*Euprymna scolopes*). Bobtail squids display bioluminescence; their light-producing organ, inhabited by luminescent symbiotic bacteria, contains reflective plates that direct the produced light in a way that prevents the squid from casting a shadow or having a visible shape on moonlit nights. The ability to add light through bioluminescence or fluorescence enables the animal to better match the actual brightness of its background.

« The red and violet colors of this yellow-edged lyretail create a strong contrast at short range since they are widely separated in the spectrum. At mid–long distance, these colors mix additively with the background, an example that resembles pointillism. Eilat, Israel.

‹ Parrotfish. Eilat, Israel.

‹ Counter-shading (Thayer's law) occurs when light illuminates the dark upper side from above making it look lighter and the underside look darker, as on this sea turtle. Bunaken, Indonesia.

>> An ornate ghost pipefish. Puerto Galera, Philippines.

> The iridescent cornea of the crown toby fish, with blue or green coloration, obscures the pupil, enabling it to merge with the body pattern. Eilat, Israel.

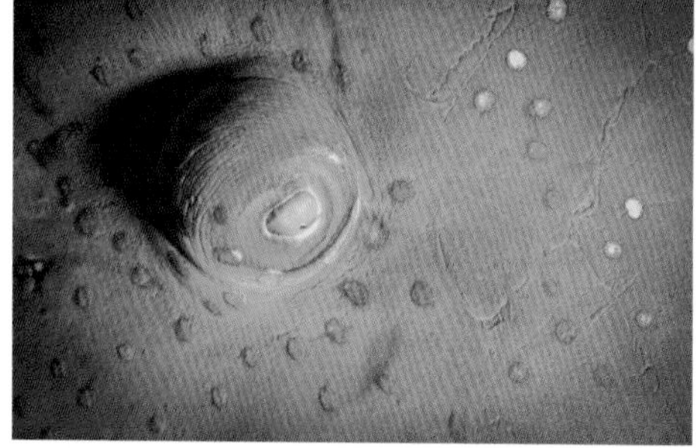

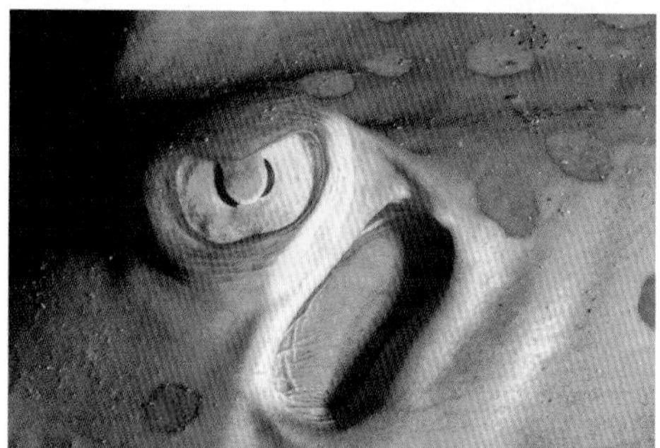

>> Spotted-ribbontail ray. Eilat, Israel.

> Broomtail wrasse eye. Eilat, Israel.

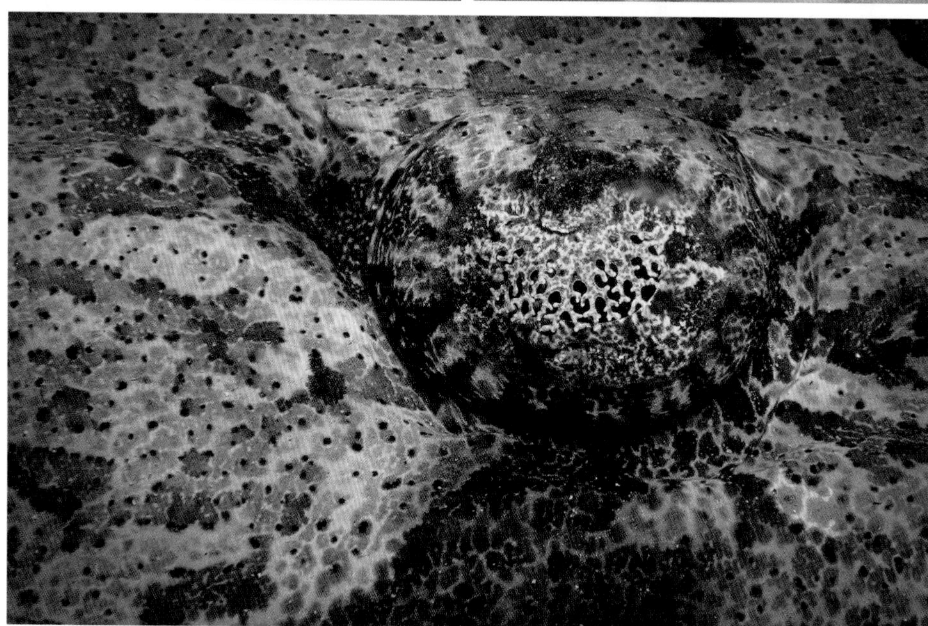

> Predators and prey commonly focus on the eyes of other fishes. To avoid fatal wounds, many fishes have developed color patterns of false eyespots on non-critical body parts. Others, such as the crocodilefish, developed well-camouflaged eyes. Crocodilefish eye. Eilat Israel.

< False eyespots of juvenile yellow boxfish. Lembeh Strait, Indonesia.

Frogfish camouflage

Frogfishes are masters of disguise and provide an excellent example of aggressive mimicry. There are approximately 50 species of frogfishes, widely spread throughout both tropical and subtropical waters; many of them can be found in the Coral Triangle. They have an unusual body shape, and the ability to change color and skin textures in order to disguise themselves. Frogfishes mimic and merge with the pattern of their surroundings, whether it is algae, seagrass, sponge or coral.

In general, frogfishes switch back and forth between two major color categories: light phase and dark phase. The light colors are often yellow or yellow-brown, while the darker ones are frequently green, black or dark red. For example, the hairy frogfish (*Antennarius striatus*) changes its color between four distinct color phases: a green phase (resembles algae) and white, orange and black phases (according to similarly colored sponges). The color change takes place over a period of days or weeks. In addition, the frogfish body is camouflaged with warts, small brown spots, patches of algae, hair-like small skin extensions known as dermal spinules, skin flaps, pink blotches and more.

^ Giant frogfish.
Puerto Galera, Philippines.

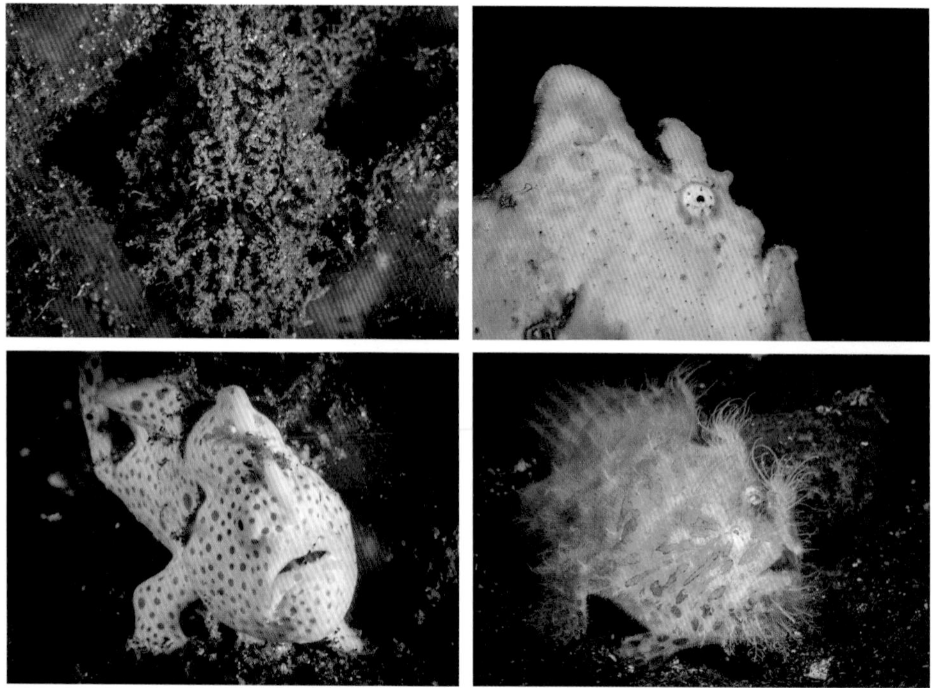

>> Giant frogfish, yellow phase. Eilat, Israel.

> Perfectly camouflaged freckled (scarlet) frogfish. Lembeh Strait, Indonesia.

>> Hairy (striated) frogfish. Lembeh Strait, Indonesia.

> Juvenile painted frogfish. Lembeh Strait, Indonesia.

Color change in pygmy seahorses

Pygmy seahorses are a group of small fishes, between 1.3 and 2.5 cm in length, most of whom live in the Coral Triangle. Some pygmy seahorses live in obligatory association with sea fans. These tiny fishes are so perfectly camouflaged that they were only discovered in 1969, quite by accident, after a pair of them were found attached to a gorgonian sea fan collected by a scientist, who took the sea fan to a museum.

Their bodies are well camouflaged, covered with rounded bumps that look very similar to the polyps on a sea fan. They feed on tiny crustaceans including juvenile crabs, lobsters and shrimps, as well as on the tissue of sea fans and the tiny zooplankton trapped in their polyps.

While the specific mechanism of pygmy seahorse camouflage is yet to be discovered, it seems to be primarily based upon neuronal and hormonal control. Unlike cephalopods such as octopuses, the color change in pygmy seahorses is gradual and slow. During the settlement phase, juveniles mimic the color and pattern of their gorgonian hosts rather than the color pre-encoded by their parents' genetics.[60]

⌄ Bargibant's seahorse (pygmy seahorse), a tiny seahorse less than 1 cm long, mimics the polyps of a sea fan. During its settlement phase, a juvenile mimics the color of fan coral. Lembeh Strait, Indonesia.

60 https://youtu.be/Q3CtGoqz3ww (retrieved March 6, 2021).

Human-Biased Point of View

Our perception of reef coloration is influenced by evolutionary mechanisms responsible for crafting our land-based vision system. The human eye responds to the visible portion of the solar spectrum in the range of approximately 380 nm violet light to 720 nm red light, with photopic vision peak sensitivity corresponding to about 555 nm green light.

As humans, we pay special attention to specific colors. For example, yellow reef fishes stand out to us. This special attention to yellow is a result of human spectral discrimination beyond 500 nm, with a significant overlap in mid–long wavelength photopigments (green–yellow–orange–red). Evolutionarily, this characteristic was probably selected because it helps primates discriminate yellow-orange fruit out of the foliage from a distance.

Likewise, our attention to red might result from the need to identify risks (warning coloration) and be sensitive to non-verbal social gestures. An example of this is blushing, described by Darwin as "the most peculiar and the most human of all expressions."

Humans' body size is also a factor that influences our attention towards specific animals. In a coral reef environment, the human body is relatively large. Thus, the fish we notice and pay attention to represent a relatively small amount of the reef biomass. Most of today's reef creatures are small (with an average length of 6 cm), cryptic and located in hiding places.

Furthermore, since underwater vision is limited across large distances, the resolving power of fish eyes is significantly less acute than that of humans (typically at least ten times less), making reef animals more conspicuous to us than to some marine creatures.

An additional factor that determines our color perception is derived from our impression from photos and movies taken underwater using artificial lights. This, of course, imprints in our mind a colorful realm that does not exist for human eyes in natural marine light (below a few meters deep).

In summary, what looks conspicuous to us might look much less conspicuous to most reef dwellers. The reef landscape is a result of nature's artwork where the "painters" who decide what is on the "canvas" are the predators, and not humans or underwater photographers.

< Well-camouflage soft coral crab (candy crab). Puerto Galera, Philippines.

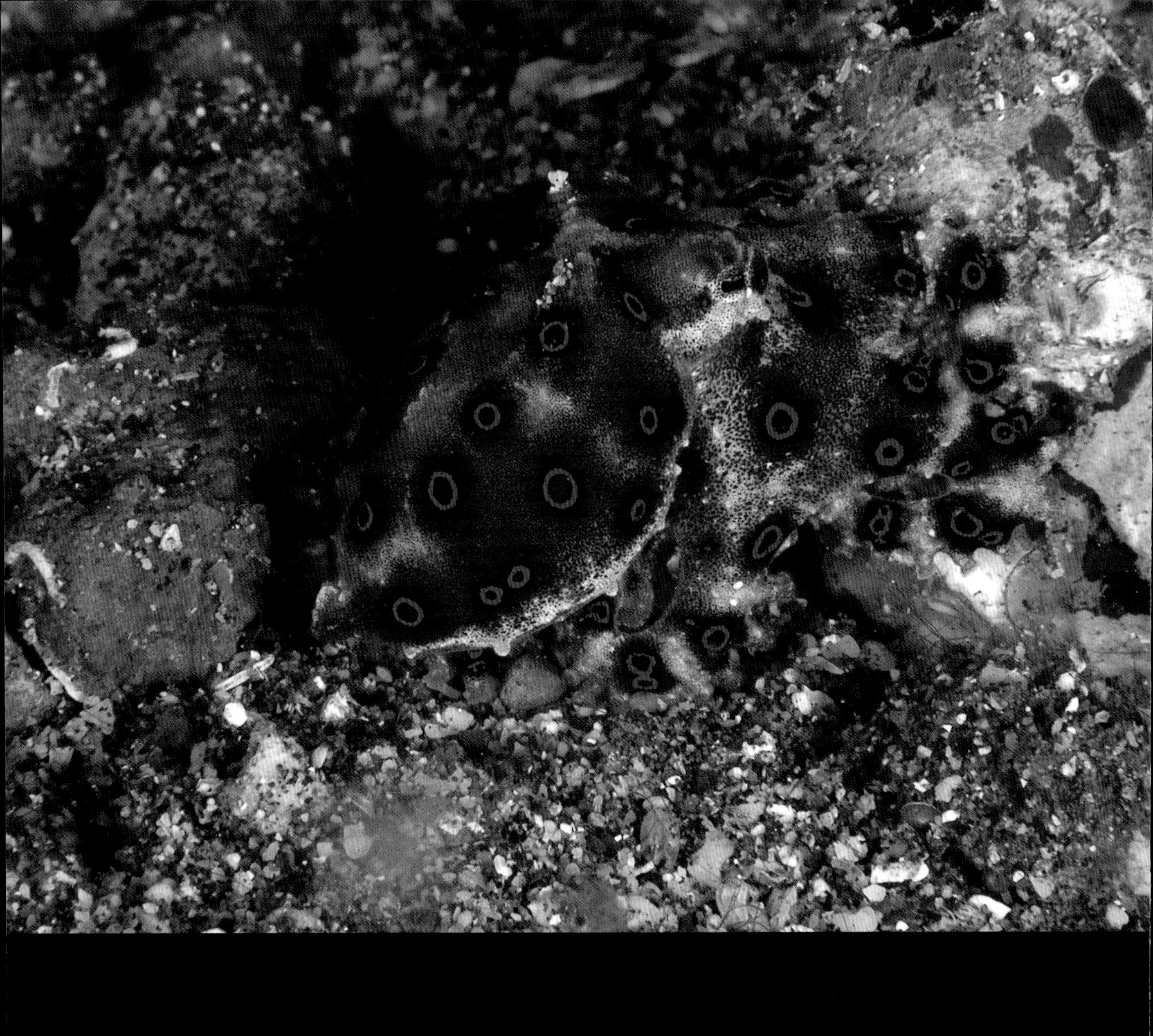

CHAPTER 4

Illumination, Disguise and Vision Mechanisms in Coral Reefs

« The blue-ringed octopus, one of the world's most venomous marine animals. Lembeh Strait, Indonesia.

The previous chapter described the primary factors that influence reef animals' coloration. In this chapter, I discuss several mechanisms of producing colors and light effects, as well as the unique vision mechanisms of several marine animals.

The first person to articulate a theory of color was Aristotle, who believed that color is a divine entity sent by God from heaven. According to Aristotle, colors are associated with the four elements that compose the universe: water, air, earth and fire. He also suggested that the sources of all colors are lightness and darkness. For almost two millennia, most scholars accepted this theory.

Today's understanding of the science of light and color began with Isaac Newton (1642–1726), who was the first to understand that white light alone comprises seven different visible colors: red, orange, yellow, green, blue, indigo and violet. Following Newton, German painter Jakob Christof Le Blon (1667–1741) was the first to define the use of three colors (red, yellow, blue) to create secondary colors (green, purple, orange). Today, we refer to Le Blon's technique as an additive and subtractive color method.

A few decades later, German poet and writer Johann Wolfgang von Goethe (1749–1832) questioned Newton's view on colors, asserting that color was not merely an objective scientific measurement, but a subjective visual phenomenon. Goethe's view was the first methodical study on the physiological effects of colors and the human visual system. Goethe attributed a psychological impact of different colors to human emotion. In his book *Theory of Colours*, first published in 1810, Goethe described red-yellow as giving "an impression of warmth and gladness, since it represents the hue of the intense glow of fire, and of the milder radiance of the setting sun." Although Goethe is mostly known for his novels and plays such as *The Sorrows of Young Werther* and *Faust*, he considered his scientific work as important as his literary achievements.

To a certain extent, and if we extend the definition of the viewer to the entire animal kingdom, Goethe was right. Today's theory of colors, which has significantly developed since the eighteenth century, stands on the shoulders of these giants (to use Newton's own metaphor) to explain his revolutionary scientific discoveries.

φ φ φ

How we see colored objects is a well-known yet complicated phenomenon. In general, the perceived color of an object is determined by the wavelength of the light reflected by it. This light is a combination of the light that hits the object and the modification mechanism on that light. In other words, the color reflectance of an object determines its color. A different way of producing light is by an internal source produced by a chemical reaction that is not dependent on external illumination.

In "standard" biological pigmentation, some incoming wavelengths (for example, the white light that includes all colors) are absorbed by the pigment that removes them from the white light combination. The color that is not absorbed is reflected to the eyes of the viewer. Thus, the absorbed colors are the ones we do not see; the colors we do see are those reflected to our eyes.

Because of light and color attenuation in water, some mechanisms of color production are more prevalent in marine creatures. In addition to standard pigmentation, the two primary mechanisms common in the coral reef are fluorescence, sometimes called biofluorescence, and bioluminescence.

Biological Luminescence Mechanisms

Fluorescence

Fluorescence night dives, known as "fluoro dives" or "glow dives," have become very popular in recent years, and provide a unique perspective of the richness of the underwater world. The special equipment required to conduct a fluoro dive is relatively simple. It requires a blue-light flashlight and a yellow filter on the mask and camera, or a UV LED, which has a wavelength just outside the spectrum visible to humans.[61]

61 The yellow shield filters out the blue light, allowing green, yellow, orange and red fluorescence to be more visible.

Science fiction author and avid diver Arthur C. Clarke (1917–2008) was among the first to experience fluorescence diving. In his 1963 sci-fi adventure novel *Dolphin Island*, written for children, one of the characters conducts a night fluorescence dive using a blue-ultraviolet flashlight.

> *When this fell upon many varieties of corals and shells, they seemed to burst into fire, blazing with fluorescent blues and golds and greens in the darkness. The invisible beam was a magic wand, revealing objects that were otherwise hidden and that could not be seen even by ordinary light.*

I experienced my first fluoro dive in Puerto Galera, Philippines. The entire reef glowed in a variety of breathtaking, fluorescent colors. Corals, fishes and invertebrates, some of which do not attract special attention during a regular dive, look like an incredible psychedelic disco when lit by the blue-UV flashlight, a phenomenon not visible to the naked human eye.

φ φ φ

At the beginning of the last century, physicists Niles Bohar (1885–1962) and Ernest Rutherford (1871–1937) presented an atom model that can be used to illustrate the phenomena of fluorescence.[62] Based on this model, in fluorescence, a photon hits an orbital electron within a specific protein molecule. The electron absorbs the light at one wavelength, often the ambient blue water light, and "jumps" to a higher, unstable energy level. For a few nanoseconds, the electron interacts with the molecular environment, while losing some energy. Shortly thereafter, the electron goes back down to the initial state, re-emitting lower energy that corresponds to a different wavelength (typically red, orange or green). In sum, the emitted color is different from the absorbed light color.

The adaptative significance of fluorescence in marine creatures is mainly attributed to the reduced efficacy of natural color pigments underwater, for both incidental and reflected light. Still, it may be that fluorescence is exploited to emit a range of colors as a response to the absorption of the ambient blue light, which is the predominant color in deep water. In other words, by using fluorescence, animals are

62 Here is a simplified version of Bohr's model: An atom is composed of a nucleus surrounded by orbiting electrons, just like planets revolve around the sun. The electrons orbit in discreet circles (like the planets in the Solar System) and cannot occupy any of the spaces between the orbits. Each orbit requires different energy.

capable of signaling in colors that are fully absorbed in deep water. This is probably the reason why fluorescence is so widespread among reef animals such as corals, jellyfishes and mantis shrimps, fishes including eels, lizardfishes, blennies, gobies and flatfishes, as well as in sharks.

Marine fishes may use fluorescence for a wide variety of purposes, including communication, predator avoidance, and perhaps even prey attraction/predation. Cryptic fishes that blend in with the reef environment perhaps use fluorescence as signaling for mating purposes and interspecies communication. It has been observed that closely related species that resemble each other emit distinct fluorescent emissions as a recognition function. Other species blend in with their surroundings under fluorescent lighting conditions and use fluorescence as a means of camouflage.

The mantis shrimp, incredible for many characteristics described in other chapters of the book, is also known for its amazing eyes and vision, which it uses for interspecies communication. Pigments in its appendages absorb the ambient blue light and re-emit it in a different color (yellow-green), resulting in typical spotty marks. The light's wavelength is distinct and can only be traced by members of its own species, which allows it to signal its fitness to potential mates, and deter other mantis shrimps from invading its space.

Another example is frogfishes that use fluorescence to attract potential prey. They attract their victims by emitting similar patterns of fluorescent light.

Bioluminescence

Bioluminescence, the light produced by chemical reactions in the bodies of living organisms, has attracted people's attention since ancient times. Although bioluminescence is much more common in the marine environment, most ancient encounters were on land, with light produced by live organisms such as fireflies. For many years, this light was akin to fire. Aristotle was the first to describe the light emitted by organisms as a "cold light" that, unlike candles, was not accompanied by heat.

There are many stories about the use of bioluminescence by ancient and modern people as a source of light. Pliny the Elder (23–79 AD), the Roman naturalist and army commander, mentioned many light-producing animals and fungi. He described a species of luminescent jellyfish that were used as torches. German botanist and physician Georg Eberhard Rumphius (1627–1702) describes how Indonesia's indigenous people use luminous fungi to find forest paths. In the nineteenth century, coal miners carried jars with fireflies as a safe light source, to avoid igniting explosive gases inside the mines.

During the Second World War, the Japanese army collected species of tiny crustaceans known as "sea-fireflies" (a species of ostracod) that produce bright blue light when wet via bioluminescence.[63] The animals were dried and ground to a powder by hand. The powder was then wetted on the battlefield, generating just enough light to read maps and correspondence, without the risk of being seen by the enemy.

φ φ φ

Bioluminescence, a widespread phenomenon across many marine species and at all depths, is sometimes produced by symbiosis with bacteria. In order to emit light through bioluminescence, two major ingredients are required: a light-emitting molecule called luciferin and an enzyme that enables the reaction, called luciferase. If you've ever noticed that water sometimes glows green or blue when you swim or hit the water, it is probably due to bioluminescence produced by single-celled marine plankton. Fishes that inhabit deep oceanic environments, where external light is limited or completely absent, use bioluminescence to produce light internally, without the use of external light sources. The absence of external light is what makes bioluminescence common in deep-water animals.

In fishes, bioluminescence is used for illuminating and attracting prey, communication, counter-illumination and territorial behavior. Bacterial bioluminescence is present in certain species of squids (*Photololigo*), octopuses (*Cistopus*), flashlight fishes (*Photoblepharon*), and golden sweepers (*Parapriacanthus*). Unlike many animals which produce light internally, golden sweepers demonstrate autogenic bioluminescence, obtaining the required ingredients from their prey. They feed on ostracods, also known as seed shrimps, which contain the bioluminescent ingredients, instead of producing it themselves.

63 Ostracods are a class of small crustaceans that can live in the open water as part of the zooplankton, or in the upper layer of the seafloor as part of the benthos.

> In some habitats, golden sweepers such as these obtain ingredients for bioluminescence from their prey. Eilat, Israel

How does the blue-ringed octopus flash its rings?

The blue-ringed octopus (*Hapalochlaena*) is one of the world's most beautiful and deadly animals. There are around ten species of blue-ringed octopuses, all of them in Asian Pacific waters. When harassed, these octopuses flash about 60 iridescent blue rings. If the octopus continues to feel threatened, it will bite, delivering a deadly neurotoxic venom called tetrodotoxin, also known as TTX.[64] The rings contain multilayer reflectors, organized to reflect blue-green light that can be seen from different angles. The fast iridescence flash is exposed by activating muscles (contraction/relaxation) located above and below the iridophore (cells).[65] Unlike other octopuses, the blue-ringed octopus does not have chromatophores on its outside.

Conspicuous iridescent patterns can be remarkably brighter and more saturated than pigmented colors, therefore providing maximum visibility. Furthermore, the peak reflectance of the blue rings is located within the range of mid- to long-wavelength sensitive opsins of many marine predators.[66] Scientists suggest that this unique mechanism evolved to create fast and conspicuous warning signals, as well as intraspecific communication.

64 TTX causes respiratory failure and motor paralysis in humans that eventually leads to cardiac arrest. So far, no antidote to it has been found.

65 Iridophore are multilayer reflector cells that change how light is reflected from the body. It is worth mentioning that some squids perform remarkable camouflage and signaling to members of their own species by using similar types of iridophore cells.

66 Opsins are a group of light-sensitive proteins found in the retina's photoreceptor cells.

Electric disco clam

An uncommon method of luminescence can be observed in electric disco clams (*Ctenoides ales*), where ambient light is scattered from tiny photonic reflective structures (nanostructures) that are positioned on the outer edge of the clam's mantle. These structures often use materials with high refractive indices compared to the substrate, which make them great reflectors. The red mantle alternately exposes the reflective structures, thereby creating a flashing effect. The disco clam is probably the first known clam that behaviorally influences its photic display. It has been proposed that flashing is part of the disco clam's predator deterrence mechanism, since there is an increase in flash rate when disco clams are threatened by predators, and the presence of sulfur in their tissue, which might indicate the use of flashing as a signal of distastefulness.[67]

⌄ Electric (disco) clam. The flashing is probably part of its predator deterrence mechanism. Verde Island, Philippines.

67 https://www.livescience.com/49312-why-disco-clams-flash.html (retrieved March 6, 2021).

Reef Dwellers' Vision System and Methods of Illumination

While the vision system of reef creatures is important, it is not necessarily their most important sense. Seawater delivers sounds and pressures much farther than it delivers light, so sensing, mapping and navigating in the sea might be much more effective using organs that sense sound and pressure, at least at larger scales. Fishes have senses that humans do not have. They can sense vibrations, pressure, sounds and in some species, electric fields. They use specialized organs called chemoreceptors to smell and taste chemicals in the water. Some of these sensory organs can sense and identify prey, predators or mates across vast distances at a magnitude much greater than these creatures can see. Thus, questions about reef creature coloration reveal an anthropocentric way of looking at the sea, giving excessive importance to vision, which stems from our own dependence upon it as humans.

Coral reef dwellers exhibit a wide variety of vision systems, which differ from each other in several aspects. The first aspect is eye type. Fishes and octopuses, for example, have camera-like eyes;[68] shrimps have compound eyes;[69] scallops have mirror-like eyes; nudibranchs and worms have various light-sensing organs that can be viewed as primitive vision systems. The second difference is visual spectrum. Many reef creatures can sense various ranges of light spectrum (mainly ultraviolet and near-infrared) and light characteristics (polarized light) that are not visible to the human eye. The third aspect refers to the processing of light signals. Like humans, some creatures heavily process light information. For example, both octopuses and humans use a third of their brain's processing power to process information from the eyes. Other animals, for example mantis shrimps, are hard-wired to discriminate colors and do not rely on processing mechanisms that enable them to interpret light signals better.[70]

Every color an animal sees is dependent upon wavelength scattering and absorption in sensory cells known as photoreceptors,[71] and no less important, the

68 Camera-type eyes such as those of squids and humans work with a single lens focusing light onto the retina, which is full of photoreceptors that convert the photon energy into electric signals sent and interpreted by the brain. Camera-type eyes generate massive amounts of signals that must be processed by the brain, a challenging task that consumes significant brain processing power.

69 Compound eyes, the most common forms of eyes on Earth, include separate units, each with its own lens and photoreceptors. Each lens operates as a single independent pixel, and therefore the maximum resolution depends upon the number of lenses in a particular area. This is why mantis shrimps and other crustaceans have compound eyes that cover a significant portion of their head surface.

70 Processing can be metaphorically compared to software, while hard-wiring can be compared to hardware.

71 Photoreceptors are special sensory cells in the retina that convert incoming light properties into electric signals that are carried to the brain, where they are interpreted. The two types of photoreceptors in the human eye are

interpretation of subsequent signals in the brain. This combination is known as *umwelt*, which means perception through the eyes of the beholder.[72] Therefore, it makes sense to assume that every color may look different to every creature.

Many animals have color vision that outperforms that of humans. Humans have three types of color photoreceptors: red, green and blue (RGB). Each of these photoreceptors detects light in a different wavelength range: red is detected at long wavelengths, green at medium wavelengths and blue at short wavelengths. Some reef animals possess four or more color photoreceptors; the mantis shrimp, whose eyes contain more than 12 color receptors(!), is considered to hold the record. With its receptors, the mantis shrimp can see the near-ultraviolet light spectrum, with wavelengths in the range of 300–720 nm (humans cannot see wavelengths shorter than about 380 nm).

Two groups of marine invertebrates that use sets of signals based on polarized light are stomatopods such as mantis shrimps, and cephalopods such as octopuses, squids and cuttlefishes.[73] Polarized light is a non-color character of light, where light waves vibrate in a single plane instead of vibrating in perpendicular planes with respect to the direction of propagation. It serves many functions in marine environments, enabling animals to communicate, detect prey, navigate, orient their bodies, and detect the shore. As depth increases, the light spectrum from the surface narrows, making reflective colors useless. However, at greater depth, a weak, constant, linearly polarized light field is present, enabling creatures to make use of it and, for instance, enhance the contrast and increase the prey detection range, including transparent prey such as zooplankton.

The position of the eyes is another factor that has a vital role in an animal's vision. Having eyes on the side, as do many fishes, is essential for creating a broad field of view, but bad for depth perception and accurately assessing distances in front.

---

rods and cones. Rods are sensitive to light level changes, shape and movement, and responsible for night vision. Cones are most sensitive to color.

72 The term "polarized light" was coined by Baltic-German biologist Jakob von Uexküll, who asserted that different organisms could have different *umwelt* (self-centered world), even though they share the same environment, since they perceive the world differently through their senses.

73 Most sources of natural and artificial light are unpolarized because they consist of a random mixture of waves that have different wavelength, spatial and polarization state characteristics. Polarized light waves are light waves in which the electric field's vibration is well defined and restricted to a single plane. Light is unpolarized if the direction of this electric field vibrates randomly. Lasers are an example of polarized light, while the sun and LED spotlights are examples of unpolarized light. The adaptive value of polarized sensitive vision can be demonstrated with a polarization filter in a camera, which increases contrast and saturation, and suppresses glare from a lake or sea surface, for example.

The Incredible Mystery of the Mantis Shrimp's Eye

The mantis shrimp is an animal with incredibly diversified skills. There are more than 450 species of them, found mostly in tropical and temperate waters. By the way, mantis shrimps are not "real" shrimps; they do not belong to the Decapoda order, but are actually part of the Stomatopoda order, which diverged from the main crustaceans branch around 350 million years ago and have since developed unique characteristics. Out of all marine invertebrates, cephalopods and stomatopods stand out for their complex nerve systems.[74] Some mantis shrimp species are known for their extraordinary eyes, as well as for the powerful claws that they use to attack prey by spearing or smashing.

Mantis shrimps have a compound eye which consists of many tiny independent photoreception units. The compound eye is the most common vision system on Earth, used by insects, spiders, crustaceans, and many other animals. Each photoreception unit behaves like a single pixel, which means that an image's maximum resolution depends on the number of photoreception units. For comparison, for a mantis shrimp's eye to have the resolution that a human eye has, it would have to have a diameter of about 15 meters. Indeed, the mantis shrimp's hexagonal packed eyes cover most of its head, maximizing the resolution per given surface.[75] Compound eyes are characterized by low resolution compared to human eyes; however, because they do not depend on the brain's visual processing or a central nervous system, mantis shrimps can react extremely quickly to visual stimuli.

Mantis shrimps have two eyes that can move independently of each other, and rotate in all directions: up-down, side-to-side, and torsionally. The compound eyes of some species possess 16–21 different photoreceptors, the largest number of photoreceptors known in any animal. The eye of a mantis shrimp might include 12-channel color vision (ranging from deep ultraviolet to far-red, 300–720 nm), 2-channel linear polarization detection, and the ability in some species to detect circularly polarized light.

74 Cephalopods, from the Greek for "head-foot," are part of the phylum Mollusca. They are a small group of highly intelligent animals that live exclusively in marine habitats. The most familiar representatives of this class are octopuses, squids, cuttlefishes and chambered nautiluses. Their most notable characteristic is that they have eight to ten tentacles or arms (a nautilus may have up to 90).

75 Hexagonal patterns are prevalent in nature because of their high packing efficiency. These patterns can be found in coral, bubble rafts, inside beehives, and more. The hemispheres of the mantis shrimp's eyes feature hexagonal packing, while the mid-band is formed from rectangular facet lenses.

The capacity to sense polarized light confers the mantis shrimp with evolutionary advantages in several areas. First, the light reflecting from an object contains a polarized component that can be received by a specific photoreceptor, and therefore reveal an animal that otherwise blends into the background (contrast enhancement). Second, mantis shrimps use polarized light to facilitate interspecies communication such as mating or territorial defense. They have evolved communication systems based on displays of polarized light patterns produced by their body organs (antennal scale). Third, it has been suggested that mantis shrimps rely on the sun and celestial polarization patterns when orienting back to their home burrows.

Many scientists believe that between four and seven photoreceptors are sufficient to encode the entire visible spectrum; three photoreceptors, like humans have, plus the ultraviolet range and the red end. In humans, the combination of the eyes and brain processing power enables us to perceive millions of different colors created by processing and comparing the activity of our three-color photoreceptors. The question is why does the mantis shrimp need so many photoreceptors, especially in light of the fact that the resolution of compound eyes is relatively low and produces crude images, at least compared to those produced by human eyes?

The immediate answer to this is that, contrary to popular misconception, evolution is far from perfect and does not always provide the best solution. Evolution is subject to available tools and resources. In other words, evolution seems to be more tinkering and less engineering. Once the basic elements of the compound eye had evolved, the evolutionary trajectory was based on increasing the number (maximizing coverage) and variety (more colors) of the photoreception units.

The mantis shrimp probably uses photoreceptors to recognize colors rather than discriminate between them. The human vision system is based on comparing different signals from each of the three-color photoreceptors, which enables us to distinguish (interpret) between millions of different colors based on the output of just three photoreceptors. However, this does consume a significant amount of brain processing power. The mantis shrimp's vision system works differently; instead of relying on the brain, it has more photoreceptors for each color, with limited capability to differentiate between various colors.[76] Although this vision system cannot discriminate between closely positioned wavelengths, it would enable the mantis shrimp to distinguish colors quickly and reliably, without the processing delay from a central brain that is required for multispectral color discrimination. This hard-wired,

76 Mantis shrimps can discriminate between wavelengths that are around 25 nm apart (the difference between orange and red) while humans can discriminate between colors that are just 1–4 nm apart.

^ The eyes of a peacock mantis shrimp. Mantis shrimps are famous for their remarkable visual systems. Their array of 16 types of photoreceptors provides complex color reception, as well as linear and circular polarization sensitivity. Each mantis shrimp eye can move independently, and exhibits a large angular range of movement, including up-down, side-to-side and torsional movements. Lembeh Strait, Indonesia.

distributed processing character might be critical to survival in hostile environments that require quick decisions.

In sum, the compound eye of the mantis shrimp versus the human eye embodies a trade-off between distributed processing in the eye that enables supreme performance, and central processing in the brain that provides better resolution and color discrimination. In a way, this compromise resembles the trade-off between hardware and software-oriented products. If you are reading this book on any electronic device, you are probably reading it on an embedded system platform composed of an electronic hardware device (such as a memory and a processor) aimed at performing a fixed specific purpose, often in real time. On top of this system, there is software (such as a book reader) that is much more flexible (for example, it can be upgraded), but is also slower. As an example, a traditional television (contrary to a "smart" television) is more hardware-oriented, and less flexible than your home computer. On the vector of embedded system software, the mantis shrimp's eye is more of an embedded, hardware type of product compared to the human eye, which is more of a software product. Both components exist in both types of eyes, in different weights.

Octopuses: The Colorblind Masters of Disguise

From the monstrous octopus in Jules Verne's *Twenty Thousand Leagues Under the Sea: A World Tour Underwater* to the octopus goddesses in Polynesian mythologies, it is clear that octopuses are a mysterious, awe-inspiring animal in many cultures.

Octopuses, with an evolutionary trajectory of hundreds of millions of years, are extremely intelligent and resourceful. They are invertebrates, a member of the phylum Mollusca, along with sea slugs and clams, part of the Cephalopoda class, which also includes cuttlefishes and squids. Octopuses have three hearts; two are solely responsible for pumping blood to each gill and one pumps blood through the body. They can move quickly via a water-jet powered mechanism and, similar to starfishes, can regenerate lost arms. The vast majority of octopuses can spread a cloud of ink and disappear within seconds into a coin-sized crack. With no bones except for a sharp beak, they can fit through almost any hole or crevice.

Contrary to what many people think, chameleons are not the masters of camouflage; this accolade falls to cephalopods. Their camouflage versatility is considered to be the most developed in the animal kingdom. In less than one second, an octopus can change its appearance with a myriad of dazzling colors and patterns, modifying its shape, color and skin texture.[77]

The color change mechanism in octopuses is unique within the animal kingdom. Basically, color change is achieved by manipulating, modifying, moving or producing pigments and reflecting substances in dedicated cells called chromatophores,[78] which differ in structure and internal mechanism from one animal to another. Animals' color-changing mechanisms can be classified into two major categories: morphological and physiological, which are characterized by different mechanisms and color-changing speeds. The morphological color change is a long-term process, resulting from changes in the morphology and density of the chromatophores. In contrast, physiological color change is a more rapid process, one that occurs in milliseconds to hours, and is accomplished by mobilizing pigments within the chromatophores.

Pigment-containing cells embedded in cephalopods' skin are novel evolutionary structures with special morphology that is distinct from those found in any other

77 Roger Hanlon's incredible underwater video clip shows an octopus's amazing camouflage capabilities: https://www.youtube.com/watch?v=JSq8nghQZqA (retrieved March 6, 2021).

78 Chromatophores are pigment-containing, light-reflecting cells present in the deeper layer of skin of various animals.

animals. They are composed of an elastic sac with pigment granules in different colors (black, orange, red and yellow), and surrounded by muscles, whose size is controlled by the nervous system. For example, if cephalopods want to resemble a dark background, black chromatophores need to spread and yellow ones need to shrink. This is significantly different from the mechanism in fishes, as well as many other animals, in that the shape of the pigment's sac changes, rather than a translocation of pigment vesicles within the cell.

This fast and dynamic change depends on three major factors: the eyes, the central nervous system, and the pigment-containing cells embedded in the cephalopod's skin. Several studies argue that pigment-containing cells might also be controlled by the local peripherally distributed nervous system. Octopuses have the ability to sense light through the skin and produce a response that changes their skin color, even when the skin is separated from the body. The skin is probably capable of gathering information about the surrounding light using the same group of proteins (opsins) that react to light in the octopus's eyes.

The longstanding puzzle of cephalopod vision

The holy grail of cephalopod coloration research is to answer the question of why cephalopods that often use bright colors as mating displays and accurately mimic colors to blend into their surroundings are technically colorblind. This question stems from the fact that, unlike many animal eyes that contain multiple types of color photoreceptors, cephalopods have only one type of light photoreceptor; this means that their worldview is monochromatic, they can see only black, white and grayscale intensities. So two major questions arise: How can they match colors to their surroundings if they cannot see? And why do they break their perfect camouflage in order to produce colorful, visible mating displays? These questions are still open and the only compelling hypothesis that has been suggested is that of Harvard astrophysicist Christopher Stubbs and his son, who proposed that cephalopods are not colorblind but use a combination of a unique optical effect called chromatic aberration and an off-axis pupil to sense colors with only one photoreceptor.

In human eyes, when light passes through a lens or prism, each color of light refracts at a different angle, and therefore to a different distance, theoretically requiring us to focus separately on each color. Cephalopods may use this phenomenon and turn it into a color vision mechanism. Instead of adjusting the lens and shifting the focus between objects at different distances (as we do when we use a microscope to view colors that were dispersed by a prism, for example) the off-axis pupil structure of the cephalopod might cause its eye to shift focus between different wavelengths (colors) of light. It is suggested that cephalopods can determine an object's color based on when its image goes in and out of focus. Unlike humans and other mammals that have round, small, on-axis pupils (that enable better acuity), cephalopod pupils are structured in a wide, off-axis u-shape, w-shape, or dumbbell-shape, a structure that enables light to penetrate the eye from many directions. When light penetrates the eye through a wide pupil, it is not focused on a single spot, as in humans, but is spread out; for this reason, it may be that cephalopods can shift focus between different wavelengths. Such a mechanism would also have drawbacks, since the vision has less acuity and consumes a lot of brainpower (processing power to analyze incoming light).

This unique mechanism for "seeing" color demonstrates the importance of brain processing in vision mechanisms. In other words, a specific color photoreceptor may not be needed in order to see color, as there may be other mechanisms that enable the brain to interpret color that is emitted or reflected from an object.

⌄ Chromatic aberration is an optical effect that occurs when a lens fails to focus on all colors at the same point. It is believed that some cephalopods, which are technically colorblind, use this effect to compensate for their lack of color photoreceptors. Their wide off-axis pupils spread the light out, enabling the cephalopods to shift focus between various wavelengths and discriminate between colors based on the focal point.

Chromatic aberration effect

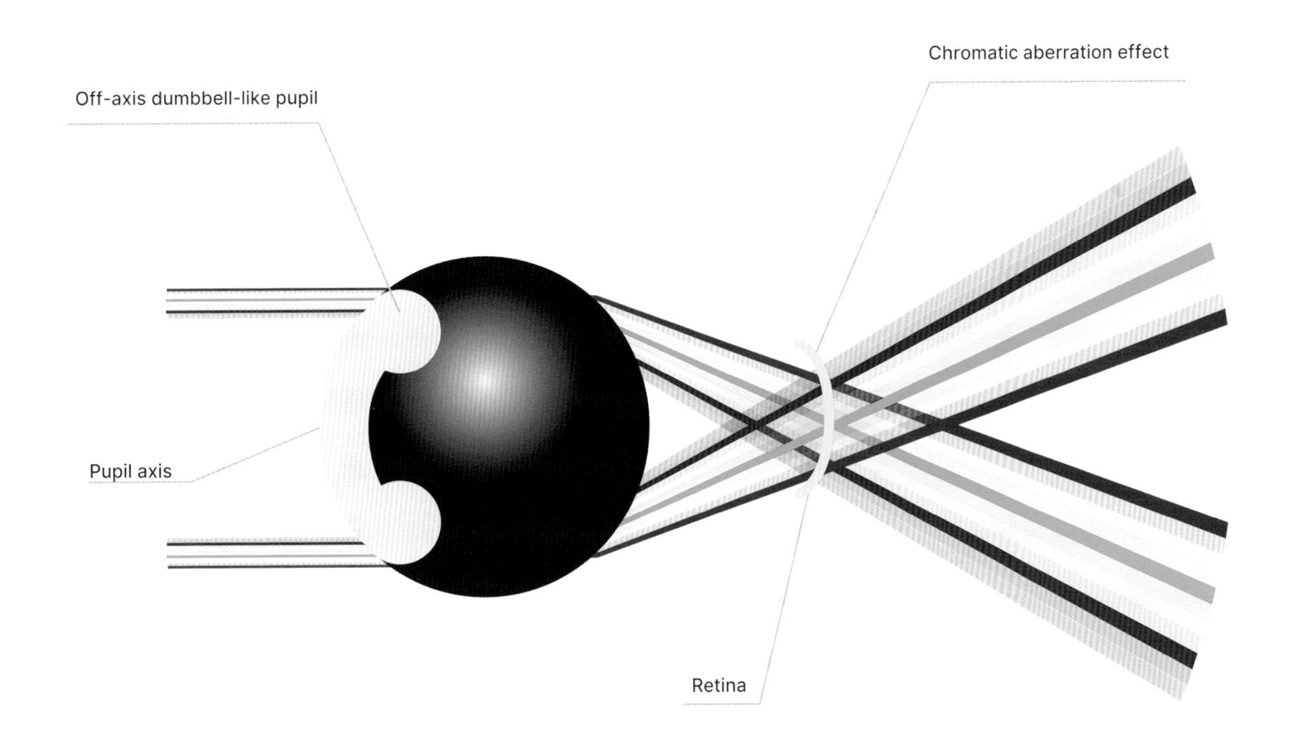

> Wunderpus octopus. Lembeh Strait, Indonesia.

v A broadclub cuttlefish near a fire urchin and sea squirt. It was suggested that the cuttlefish's wide, off-axis w-shaped pupils enable it to sense colors with just one light photoreceptor using a vision mechanism known as chromatic aberration. The eyeless sea urchins in the background also have a special vision mechanism; they can sense light via light-sensitive cells that are spread across the entire surface of their bodies. The spines are used to break up the light and pick out relatively fine visual detail. Puerto Galera, Philippines.

^ The flamboyant cuttlefish, a master of camouflage, can quickly change to a pattern of dark brown, black, white and yellow patches, and may show bright red on the arm tips. Puerto Galera, Philippines.

"Life did not take over the world by combat, but by networking."

Lynn Margulis and Dorion Sagan, *Microcosmos: Four Billion Years of Microbial Evolution*, 1997

CHAPTER 5

Mutual Aid
Coral Reefs as a Symbiotic Society

« Mutualistic symbiosis between a sea turtle and suckerfishes. In ancient times, people believed suckerfishes had magical powers that could stop ships from launching. Bunaken, Indonesia.

Prince Pyotr Kropotkin (1842–1921) was a Russian polymath, mostly known for his political thoughts and advocacy of anarcho-communism.[79] Kropotkin, an aristocratic graduate of an elite Russian military academy, represented a different school of thought than Darwin regarding the theory of evolution. While Darwinism was understood, to a certain extent mistakenly, as a theory based on the tenet of the "struggle for existence" and "survival of the fittest," Kropotkin advocated that cooperation between organisms, as a means of overcoming the forces of natural selection that act on the group selection level, was underestimated. As he wrote in his 1902 collection of essays, *Mutual Aid: A Factor of Evolution*, "In all these scenes of animal life which passed before my eyes, I saw mutual aid and mutual support carried on to an extent which made me suspect in it a feature of the greatest importance for the maintenance of life, the preservation of each species, and its further evolution."

Kropotkin argued that selection for mutual aid directly benefits each individual organism in its struggle for success. He distinguished between the struggle over limited resources between organisms of the same species, which leads to competition, and their joint struggle against the hostile environment, which leads to cooperation. Kropotkin believed that mutual aid and solidarity is a universal, vague feeling or instinct that characterizes humans as well as animals. "It is not love to my neighbour–whom I often do not know at all which induces me to seize a pail of water and to rush towards his house when I see it on fire" he wrote, "it is a far wider, even though more

79 Anarcho-communism (also known as anarchist communism and libertarian communism) is a form of anarchism that supports the elimination of the state and capitalism in favor of non-hierarchical voluntary associations, based on the guiding principle similar to a slogan made popular by Karl Marx: "From each according to his ability, to each according to his needs."

vague feeling or instinct of human solidarity and sociability which moves me. So it is also with animals."

It is interesting to see how different schools of thought developed as a result of varying environmental conditions and the prevailing zeitgeist. Two major experiences led to Kropotkin's evolutionary thought. The first was his own experience as a naturalist conducting a geographical survey expedition in the wilderness of eastern Siberia in order to find the best route for what was later known as the Trans-Manchurian Railway. The second was the profound impression left upon him by the egalitarian fraternity amongst watchmakers in a cooperative in Switzerland's Jura canton which led him to anarchism and the political opinion that central government law should be removed in favor of local community consensus.[80] Today, Kropotkin is mostly known for laying the groundwork in the evolutionary study of altruism and synergy, as well as his significant contribution to the school of thought that questions the political value of competition.

In contrast is the British doctrine, developed independently by Darwin and Wallace following their experiences in the tropics, an area packed with species and the highest biodiversity on the planet. This school of thought is in line with Thomas Hobbes' (1588–1679) theory of politics, where the absence of a central government leads to "war of all against all." Economist Adam Smith (1723–1790), the father of capitalism, and clergyman and political economist Thomas Malthus (1766–1834), who had forecast a population explosion and food supply shortage, both applied the same principle to the problem of population growth and competition.

The dispute over the boundaries of an organism,[81] and cooperation, has accompanied academic thought throughout the twentieth century. However, the British doctrine was dominant until the 1960s, as reflected in the discourse about evolution that largely underestimated cooperative phenomena.

As elaborated in the section about superorganisms (Chapter 2), Lynn Margulis coined the term holobiont, an entity made of different species (bionts) that take part in symbiosis, asserting that a holobiont confers a significant increase in fitness not available to any species by itself. The theory of the holobiont and the benefits it brings to its bionts somewhat resembles the economic theory of "comparative

80 Kropotkin was so impressed by his experience in this cooperative that he wrote in *Memoirs of a Revolutionist*: "the egalitarian relations which I found in the Jura Mountains, the independence of thought and expression which I saw developing among the workers, and their unlimited devotion to the cause appealed strongly to my feelings; and when I came away from the mountains, after a week's stay with the watchmakers, my views on socialism were settled. I was an anarchist."

81 For example, do the boundaries of human organisms include the viruses, bacteria and fungi that live on and in us?

advantage" contemplated by British economist David Ricardo (1772–1823). Simplified, this principle suggests that the combined output of two countries increases if they apply the principle of comparative advantage compared to the output that would be produced if the countries choose to be self-sufficient and to allocate resources to the production of both products domestically in each country. The theory of comparative advantage advocates for international free trade, and shows that even if a country benefits from an absolute advantage in the production of goods, trade can still be beneficial to both trading partners.

What is Symbiosis?

A significant source of both evolutionary novelty and the evolution of complex systems over time is symbiosis. Symbiosis, which means "living together" in Greek, is defined as a cooperative or intimate living relationship between organisms from distinct species, with benefits to at least one of the organisms involved.

One distinction between types of symbiosis is whether the symbiosis is obligatory or facultative. Obligatory symbiosis is when one side, also known as a symbiont, depends upon the symbiosis for its survival. Facultative symbiosis is when an organism can survive without the symbiosis, although the symbiosis increases its fitness.

The term synergy also relates to the big evolutionary question regarding the unit of natural selection. Synergy is the benefit to the whole created by cooperation between symbionts, which is usually greater than the sum of its parts. It appears that natural selection acts on the individual organism, the biont, as well as on the entire "synergy unit" including more than one organism, the holobiont. The forces of natural selection that act on the holobiont provide an opportunity for greater specialization, which gives the holobiont an edge over its competitors. In many cases, these unique and beneficial relationships, although they may be beneficial to only one party, or even parasitic, end in divergence and the creation of new species complexes. It would not be an exaggeration to say that without synergy, life on Earth would be unrecognizable.

Symbiotic relationships are divided into three major groups, based on whether one symbiont is beneficial, harmful, or has no effect on the other. Mutualism is when both parties benefit from the symbiosis. Commensalism is when the symbiont utilizes the host (or the other party in the symbiosis), without benefiting or harming it. Parasitism is when a symbiont uses the host as a resource and harms it. These relationships serve two main functions: access to food and protection.

Coral reefs harbor a remarkable number of symbiotic relationships. One cannot dive in a reef without noticing one animal living alongside, inside, beside, or on top of another. Some of these animals compete for reef resources or play a role in the food web; others cooperate in various modes. In many seemingly symbiotic encounters in the coral reef, the driving forces behind the symbiosis are not completely clear, raising questions that are still unanswered, such as: How did the symbiosis evolve? What is the benefit or harm to each side?

The Centrality of Coral's Symbiosis with Algae

The tight symbiotic relationships between many corals and algae are not exclusive, and algae are central to the existence of other reef animals as well, such as species of sea slugs, sea anemones, sponges and giant clams.

Like the relationships with corals, these relationships are pretty straightforward: the algae convert sunlight into oxygen, remove wastes, and provide the host with the organic byproducts of photosynthesis, which fuel the host's growth and reproduction. The host, in turn, protects these tiny algae from predators and provides some of the essential nutrients they need to survive in the nutrient-poor ocean.

For example, the solar-powered nudibranch, a species of aeolid nudibranchs, stores algae in its cerata, a horn-shaped organ structure on the back of the aeolid that aids in respiration. They use the algae's ability to convert the sun's energy into sugar and other extra nutrients, similar to plant photosynthesis.

In a way, this fundamental relationship resembles lichens, that are also composed of a tight symbiosis between algae and fungi, and for many years were considered a single stable unit.[82] It is fair to say that given the limited supply of nutrients from the surrounding open ocean (see Chapter 2: Darwin's Paradox), it is mainly photosynthesis that provides corals with the energy to grow and build reefs.

---

82 The tight symbiosis between fungi and algae enables lichens to grow in almost all terrestrial surface habitats; on trees, rocks and houses, and in forests, deserts, tundra, etc. It is estimated that lichen-dominated vegetation covers 6 percent of the Earth's terrestrial surface.

^ Solar-powered *Phyllodesmium*, a type of sea slug, contain photosynthetic algae that enables it to draw energy from sunlight. The brown patches covering the nudibranch are clusters of the symbiotic algae. Lembeh Strait, Indonesia.

Cleaning Stations: How Did They Evolve?

Cleaning stations are very common in coral reefs and an excellent example of why nature is not a mere "red in tooth and claw" struggle for existence, but is largely based on cooperation.[83]

A cleaning station is a mutualistic, non-kin cooperation between unrelated species. Many marine species use cleaning services provided by wrasses, shrimps, and juveniles of various fishes, all usually located in specific cleaning stations. Usually, one organism (for example, wrasses) cleans another organism (for example, groupers) by removing ectoparasites from their bodies.[84] Sometimes, the cleaning is performed by the cleaner entering the gill chambers or mouth of the host, usually without ending its life in the host's stomach.

83 That famous phrase was coined by British poet Alfred Lord Tennyson (1809–1892).

84 Ectoparasites are parasites that live outside the body of the host.

The diet of some cleaners is based entirely on these parasites. Cleaners, whether shrimp or fish, are characterized by specific colors and behaviors, which have evolved independently in what is called convergent evolution.[85] Scientists have managed to demonstrate the strong selection of cleaning processes. As an example, even groupers that were grown from infancy in a lab tended to avoid eating cleaning fish that were placed with them in the same aquarium.

Most animals are incapable of removing all parasites from their own bodies, thus the benefit of allowing another animal to do so is clear. Research conducted in the Great Barrier Reef by scientists from the University of Queensland, Australia, found that without cleaners, parasite numbers increased, between dawn and sunset, about fivefold! Cleaning stations are busy places; in a separate study conducted by the same group of scientists, one cleaner fish was found to inspect more than 2,300 client fish, from over 130 different species, in a single day. Each cleaner fish eats a fantastic number of parasites per day, approximately 1,200![86]

The seemingly altruistic behavior in cleaning stations is curious since it seems to embody inherent conflict. The primary question is how did this mutually beneficial association start, evolve and imprint, even though there is an inherent conflict of interests between the associated parties? American evolutionary biologist Robert Trivers (1943–) showed that in certain conditions, altruistic behavior could be subject to selection, even between non-related species. Prior to Trivers, altruistic behavior was demonstrated and modeled mostly in the case of kin-selection; that is, animals being more altruistic towards close kin than to distant kin (for example, eusocial insects).

Symbiotic cleaning relationships, a case of reciprocal altruism, resemble, at first glance, the prisoner's dilemma:[87] cleaners can "cheat" by eating the mucus and small pieces of flesh instead of the parasites; hosts that come to be cleaned can "cheat" by eating the cleaners at the end of the cleaning session. This type of symbiosis has a time-based dependency, which means that one partner helps the other, and has to wait before it is helped in turn. For the fish coming to get cleaned, it seems that natural selection would prefer the dual benefits of getting cleaned from parasites and then eating the cleaner. Therefore, the tit for tat strategy,[88] the preferred strategy in the prisoners' dilemma, is not plausible, since the predator/cleaner might terminate the

---

85 Convergent evolution refers to unrelated organisms that independently evolve similar structures, traits and features, which originate from different ancestors.

86 http://www.cleanerfish.com/the-cleaner-fish-chronicles.html (retrieved March 6, 2021).

87 https://plato.stanford.edu/entries/prisoner-dilemma/ (retrieved March 6, 2021).

88 Tit for tat is a game-theory strategy in which each participant imitates the action of its rival after cooperating in the first round. This model, developed by Axelrod and Hamilton, provides the highest value for the participants. It is not valid in our case, however, since it assumes that the choices are made simultaneously.

game if it cheats the first time. Hence, unlike simple mutualism, where behavior does not depend upon the immediate receipt of benefit, cleaner fish are unconditionally cooperative towards those who could be their potential predators.

Due to this time-phase dependency, there are only a few, very specific circumstances in which the altruistic party can be guaranteed that the response to its altruistic act will eventually bring it benefit.

Trivers found that the preconditions for this type of reciprocal altruism are: (a) ectoparasites are unbearable; (b) finding new cleaners is difficult; (c) cleaners are site-specific and territorial, so that the same cleaner can be found and reused by the host; and (d) cleaners live long enough to be used repeatedly by the same host. Although these preconditions for the evolution of reciprocal altruism are specialized, many species meet them and demonstrate this type of altruism.

Significant field evidence supports the view of this precondition for the evolution of reciprocal altruism in the cleaner fish example. The host's altruism is explained by the evolutionary benefit of quickly and repeatedly returning to the same cleaner. Scientists have observed recurring visits of the host to the same cleaning station if the previous interaction did not lead to any significant conflict; in many cases, the hosts receive cleaning services from the same cleaners (who have an average life span of several years). From the host's perspective, it is unprofitable to eat the cleaner; if it does, it may have difficulty finding a different cleaner when it needs to be cleaned again.

Another hypothesis is based on the existence, in both humans and other vertebrates, of the rewarding effect of tactile stimulation.[89] Reef fish are no exception, and it seems that stress may be reduced by physical contact. Cleaners uses various movement gestures in order to stimulate their clients towards cooperative behavior. It has been observed that the cleaner fish influences the client's decisions by physically touching it with its fins. Moreover, predatory clients receive tactile stimulation from cleaning wrasses more often than non-predatory clients. Cleaners use tactile stimulation in a wide variety of contexts, including building trust with new clients, reconciliation after having cheated (by biting a client's mucus), and as a gesture to approach new clients. Research has shown that physical contact alone is sufficient to produce fitness-enhancing benefits, a situation so far only validated in humans, that is also supported by the positive aspect of the social relationship.

Due to the cleaners' presence and the tactile stimulation, preying instincts seem to be repressed at cleaning stations, and they are considered safe havens from

89 Tactile stimulation, in humans and other mammals, reduces stress and has therapeutic benefits, though it is unclear whether the benefits are a result of the physical contact or have social value.

predator aggression. The client remains at a standstill and is willing to tolerate occasional painful bites in exchange for tactile stimulation and relief from parasites. These two general hypotheses for the evolution of cleaning stations are not mutually exclusive, and the latter may serve as a basis for the development of the former.

Another interesting phenomenon in cleaning stations is third-party punishment, which characterizes groups and is considered adaptive on the group selection level; it is also prevalent in human societies. If a female cleaner cheats by biting a client's mucus, instead of eating the parasites, during the client's visit to the cleaning station, her male partner will punish her by chasing her. Females seem to be more cooperative following the punishment, yielding direct foraging benefits to the male.

The cleaning station is a marvelous phenomenon that serves as a platform for a better understanding of the application of game theory, with conflicting initial conditions, to the evolution of cooperative behavior in fish.

<< A sea snake (banded sea krait), with a bluestreak cleaner wrasse. Although generally not aggressive, sea snakes are venomous and can paralyze much larger animals, including humans, with a bite. Similan Island, Thailand.

<< Cleaner wrasse cleaning the inside the gills of a semicircle angelfish. Nusa Penida, Indonesia.

< Morays are frequent cleaning station visitors. Bali, Indonesia.

< My son, Or, enjoys a bit of dental cleaning by Bluestreak cleaner wrasse. Sinai Peninsula, Egypt.

The Sea Anemone and Its Partners

Sea anemones are predatory animals, related to corals and jellyfishes, and part of the phylum Cnidaria. They take the form of an array of tentacles filled with venomous stinging cells called cnidocytes;[90] each cnidocyte surrounds a single polyp which is usually attached to a substrate. When prey comes within range of their tentacles, sea anemones have a powerful stinging response, allowing them to immobilize plankton as well as small fish. Sea anemones' natural enemies include certain sea slugs, starfishes, eels and flounders.

Like many corals, some sea anemone species obtain many of their nutrients by forming a symbiotic relationship with algae; they provide the algae with protection and in return receive oxygen, sugar, and other by-products of photosynthesis. The most iconic sea anemone symbiosis is the relationship between the sea anemone and the anemonefish (clownfish). There are about 1,000 sea anemone species, and about ten of them coexist with nearly 30 species of tropical anemonefish. This symbiosis is a mutualistic symbiosis, where the stinging tentacles of the sea anemone defend the anemonefish against possible predators and in return, the anemonefish protects the sea anemone from its predators and removes some of its parasites. The anemonefish is protected from the cnidocytes by a thick external mucus. Research has shown that the anemonefish is innately protected from some anemone species, but must acclimate to live with others. Unlike sea anemones, many species of anemonefish cannot thrive or live for long periods on their own outside the protected area, and are profoundly dependent upon this relationship (obligatory symbiosis).

Anemonefish are not the only fishes with species that are immune to the stinging tentacles of sea anemones. Some species of wrasses, cardinalfishes, damselfishes and more are also adapted to sea anemone tentacle stings. Wrasses are usually free-living, but proximity to sea anemones may provide them with protection from predators. Additionally, it is likely that the wrasses' diet includes organisms that have been adversely affected or killed by sea anemones. Similarly, crabs and shrimps also often form mutually symbiotic relationships with sea anemones, again for protection from predation. For instance, the beautiful porcelain crab lives and captures its food from within the tentacles of a few species of sea anemones.

---

90 Cnidocytes are highly specialized cells containing a stinging organelle called a cnidocyst. These cells are used by cnidarians for capturing and immobilizing prey using toxins contained within the cells.

< Banggai cardinalfishes swim among the tentacles of large sea anemones. Bunaken, Indonesia.

∨ Spotted porcelain crabs are filter feeders that live among the tentacles of sea anemones. Puerto Galera, Philippines.

> Anemone shrimp and clownfish inside a sea anemone. Puerto Galera, Philippines.

> "Yawning" Red Sea anemonefish. Eilat, Israel

⌄ Anemone shrimp. Eilat, Israel.

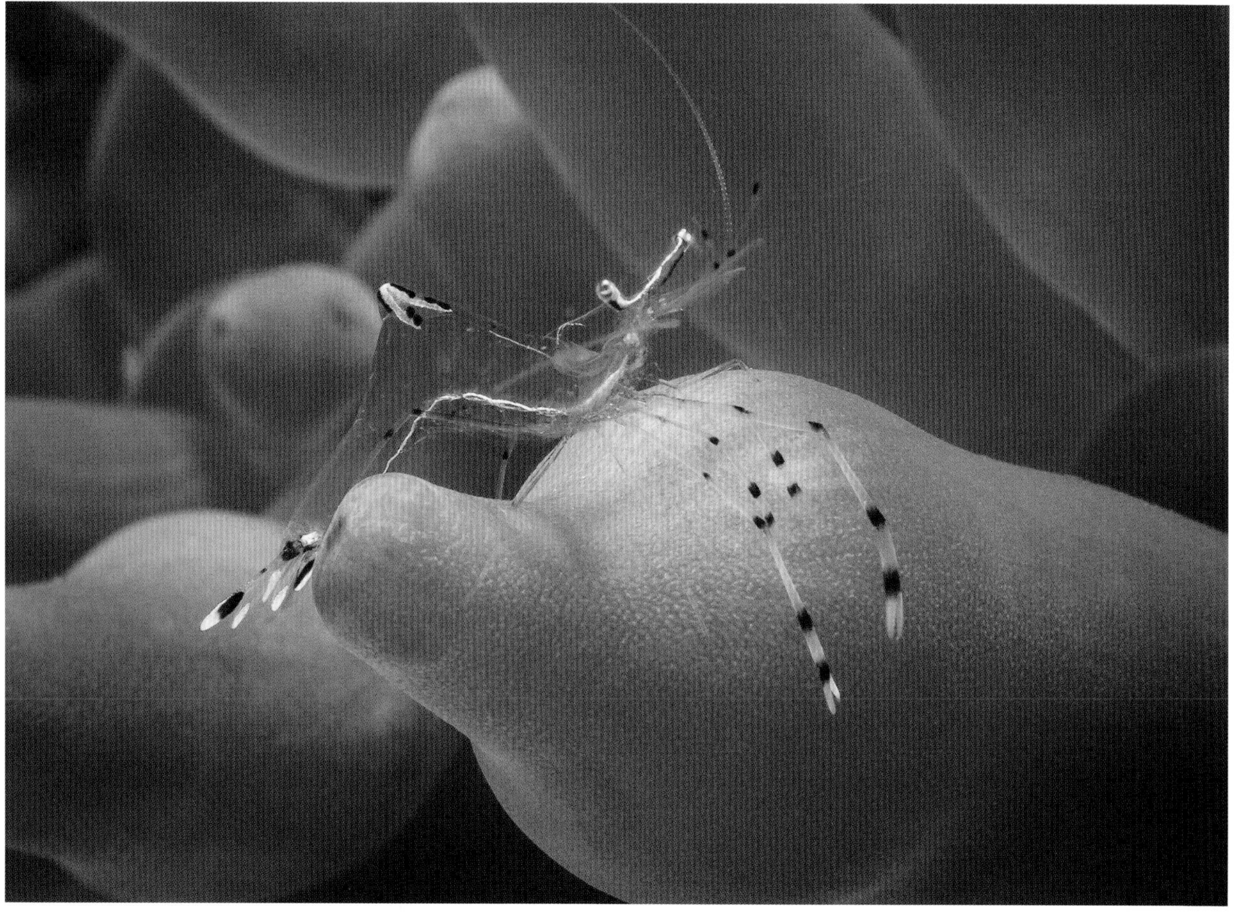

Examples of Symbiosis with Crustaceans in the Reef

Mosaic boxer crabs and sea anemones

Boxer crabs and certain species of sea anemones form one of the most unusual symbiotic relationships in the entire coral reef system. The crabs attach sea anemones to their claws and use the sea anemones' stinging tentacles as a defense tool; in exchange, the crabs feed their devoted defenders. This partnership seems to be obligatory from the crab's perspective. When a boxer crab lacks a sea anemone on its claw, it tends to steal one (either a whole sea anemone or part of one) from another boxer crab. Stealing part of the sea anemone is, from the sea anemone's perspective, a form of asexual reproduction, since eventually two genetically identical anemones grow independently on each crab. Laboratory tests show that a pair of anemones held by a particular crab are usually genetically identical. So, this is a remarkable case in which one species (the boxer crab), induces the asexual reproduction of another species (the sea anemone).

‹ Mosaic boxer crab with pom-pom sea anemones attached to its claws. Lembeh Strait, Indonesia.

Emperor shrimps – commensalism with multiple hosts

The emperor shrimp (*Zenopontonia rex*) is a species of small commensal shrimp that is part of the Palaemonidae family and found in the Indo-Pacific Ocean. Known as the "hitchhikers of the reef," these shrimps are often found on the backs of nudibranchs, sea cucumbers and many other species. Emperor shrimps are both carnivores and detritivores, which means that they feed on the diet of other animals, or what is left over from it. In some cases they have been spotted eating parasites and fungi on their hosts, which indicates that the relationship with the host could be defined as mutualism.

> A commensal emperor shrimp on a blue sea star. Lembeh Strait, Indonesia.

> An emperor shrimp on a sea cucumber. These shrimps are very common in coral reefs and associated with a large number of invertebrates such as sponges, corals, sea lilies, sea stars, sea cucumbers, mollusks and more, mainly as cleaner shrimps or commensals. Their diet includes the hosts' leftovers, waste and parasites. Lembeh Strait, Indonesia.

Alpheid shrimps and gobies

The symbiosis between pistol shrimps (snapping shrimps) and goby fishes is one of the most widespread mutualistic relationships in the coral reef. The foundation of this relationship is based on the animals' mutual use of the shelter that is burrowed, constructed and maintained by the shrimp. In return, the goby guards and protects the entrance to this burrow. The shrimp keeps in constant physical contact with the goby, touching it with its antenna, so it can sense the goby's warning signals. An additional possible benefit to the shrimp might be its consumption of the goby's excrement inside the burrow.

‹ A spotted shrimpgoby and an alpheid shrimp. The goby is on guard while the shrimp maintains its burrow (the shrimp's body is partially inside the burrow). Puerto Galera, Philippines.

Urchin carry crabs

Occasionally, divers in the Pacific Ocean encounter sea urchins that seem to be walking. A closer look reveals that the sea urchin is actually being carried on the back of a carrier crab. The sea urchin gets food from the crab as well as a lift to a new feeding area. The crab, for its part, is protected from predators by the sea urchin's sharp spines.

> Urchin carry crab.
Lembeh Strait, Indonesia.

Sponge decorator crabs

Some species of decorator crabs use sponges to protect themselves in two ways. First, by covering their shell with sponges, the crabs gain an excellent camouflage. Second, thanks to some of the sponge's materials (for example toxins), potential predators are deterred. These are non-obligatory mutualistic relationships since the sponges attached to the crab's body continue to flourish and filter feed as if they are located on the reef, and further benefit from the crab's mobility and possible additional feeding opportunities.

> Sponge decorator crab.
Lembeh Strait, Indonesia.

Other symbioses in the subphylum crustacea

Many of the shrimps, crabs, and lobsters we encounter in the reef live in various types of symbiosis with a wide variety of reef creatures. Many of these crustaceans are well camouflaged and adopt a cryptic marine lifestyle. Their symbiotic mode of life goes along with a vast array of morphological body plan changes. Crustaceans occur symbiotically with a large variety of cnidarians, but also with a large variety of other host organisms, including sponges, mollusks, echinoderms and ascidians. The symbiotic relationships can range from facultative to obligatory symbiosis, and from mutualism through limited commensalism to semi-parasitism. Some crustaceans are quite adaptable and can use a variety of species for their symbiosis; others may be highly specialized to only one host.

« Zanzibar whip coral shrimp. Lembeh Strait, Indonesia.

‹ Coleman shrimp on a fire urchin. Puerto Galera, Philippines.

« Spiny tiger shrimp. Lembeh Strait, Indonesia.

‹ Hairy squat lobster on sponge. Verde Island, Philippines.

"Who trusted God was love indeed
And love Creation's final law –
Tho' Nature, red in tooth and claw
With ravine, shriek'd against
his creed..."

Alfred, Lord Tennyson, "*In Memoriam, A.H.H.*," 1850

CHAPTER 6

Nature, Red in Tooth and Claw

Defense and Preying Mechanisms

« Reef lizardfishes ready to ambush passing fish. Eilat, Israel.

In its broad sense, predation, which can be defined as an organism preying upon another organism for dietary purposes, is as old as life on Earth, and served as a significant factor in some major evolutionary transitions. For example, it is suggested that billions of years ago, the origin of eukaryotic cells (cells containing a nucleus) was probably the result of predation among prokaryotes (cellular organisms that don't contain a nucleus). Mega predators appeared much later, during the Cambrian period (climaxing around 550 million years ago) and have since shaped the evolutionary trajectories of animals. Unlike the common belief, predation is not an advanced feeding strategy, and actually pre-dates herbivory (plant-eating). The old history of predation has been substantiated by a computer simulation suggesting that successful strategies such as predation could have evolved fairly rapidly, within a thousand generations or so. In other words, there may never have been a world without predation in evolution. Such a "Shangri-La" for primary producers and decomposers is hard to imagine, and would require very different selective pressures from those that affect most animals today.

Throughout history, humans have used two primary methods for making sculptures: carving and modeling. Similarly, these two methods also stand behind the shaping of organisms' fitness and the creation of complex living organisms. In subtraction (carving), which characterizes artists such as Michelangelo di Lodovico Buonarroti (1475–1564), the artist starts with a block of material such as marble and uses a hammer and chisel to remove pieces until the statue is complete. As Michelangelo reportedly said, "In every block of marble I see a statue as plain as though it stood before me, shaped and perfect in attitude and action. I have only to hew away the rough walls that imprison the lovely apparition to reveal it to the other eyes as mine see it."

The second method, used by sculptors such as Auguste Rodin (1840–1917), is known as addition (modeling). In this method, you start with a lump of clay and add to it until you get the final clay model, and then make a plaster cast, which enables you to cast the bronze.

The evolution of predation mechanisms works similarly to the above sculpturing methods. Natural forces are like subtraction; they are capable of weeding out freaks, failures and less fit traits. A mutation is like an addition; it can add traits that sometimes increase fitness. Together, these two processes are responsible for the complexity of evolutionary biology. One way that these processes can take place is an evolutionary arms race. An adaptation in one species influences adaptations in other species, giving rise to counter-adaptation and so on, playing a significant role throughout the co-evolution of both species. A prominent example of an arms race that occurred during the Cambrian period is the development of a shell. In a relatively short period of time, many animals developed a shell, which is energetically costly and would not have evolved without the need to defend against the spread of predators.

There are arms races between predators and prey, parasites and hosts, pathogens and antibiotics, and even between male and female members of the same species. The phenomenon can be summed up in a phrase attributed to Greek philosopher Heraclitus, active around 500 BCE: "All things come into being by conflict of opposites, and the sum of things flows like a stream."[91]

φ φ φ

Broadly speaking, an arms race can fall into one of two categories, symmetrical and asymmetrical. An asymmetrical arms race occurs when natural selection acts in the same direction, and species adopt the same way of coping with the selection force. An example of this is the competition between various coral species for the sun's radiation. An asymmetrical arms race is the result of contrasting forces where the species adapt in different ways. An example of this could be the evolutionary arms race between a predator and prey. For instance, many mollusks have evolved thick shells and spines to avoid being eaten by animals such as crabs and fish, which in turn developed powerful claws or other mechanisms that allow them to successfully prey on the mollusk, despite its armor.

91 Heraclitus was quoted by Diogenes Laërtius, the biographer of Greek philosophers, in his book *Lives of Eminent Philosophers*. The original work was probably first compiled in the third century BCE.

Arms races can evolve gradually and through a positive feedback loop,[92] which usually ends with the appearance of internal or external constraints that bring the system into equilibrium for a certain period of time. Some of the phenomena described in this chapter are so extreme that it makes sense to assume that they result from an evolutionary arms race.

The various tasks performed by these creatures require biological materials that have developed during hundreds of million years of evolution. They are mostly complex composites, whose mechanical features are often remarkable, considering the weak elements from which they are created. Unlike synthetic systems, these biological materials are characterized by hierarchy, self-organization, multifunctionality, self-healing capabilities (for example, regeneration), and the ability to be repeatedly fueled by a metabolic process. For example, the ability to self-heal is nearly universal in nature, and following trauma or injury, most structures can repair themselves throughout the life of the organism.

Biological ultrafast movements are an excellent example of nature's superiority over human-made systems. Biological systems are smaller in scale, can be used repeatably with high fidelity, and are more efficient, robust and lightweight. These astonishing systems optimize the universal trade-off between force and velocity, in a challenging environment and across a variety of size scales.

In general, animals accomplish swift actions by using mechanisms that decrease the duration of the movement. This process, known as power amplification, is composed of three functional units: an "engine" (for example, a muscle) that slowly loads an "amplifier" (for example, a spring that stores energy), which rapidly moves a "tool." A demonstration of two of these unique, ultrafast biological systems is described in the following sections.

Cnidaria Preying Mechanism

Species of the phylum Cnidaria are characterized by a simple nervous system and a stinging cell called a cnidocyte (or nematocyte). The stinging cell contains a central capsule-like organelle known as a cnidocyst (or nematocyst), which is responsible for delivering the sting and immobilizing the prey. Cnidocytes are highly

92 A positive feedback loop is a process in which the end products of an action reinforce an increase in that action (when A produces more of B, which produces more of A and so forth). In contrast to negative feedback loops, positive feedback loops intensify change and tend to move a system away from its initial state. Positive feedback loops have driven many of the most significant changes in evolution and ecology.

specialized cells that are a common feature of cnidarians, including corals, jellyfishes, sea anemones, hydrozoans and more. They are mostly used for prey capture and defense from predators.

Many stinging cells include an oval-shaped capsule to which a long barbed thread is attached. The thread is tightly coiled within the capsule and, in some stinging capsules, might be armed with spines and toxins. It is expelled independently or triggered by supporting sensory cells from within the capsule in a harpoon-like release, similar to a jack-in-the-box. In order to prevent wasting this "silver-bullet" style weapon, sensory cells surround the stinging cell, helping to regulate the capsule's discharge by reacting to a correlated combination of stimuli, including vibration, light and mechanical and chemical changes.

The explosive discharge of the capsule within the stinging cell, often accompanied by the release of toxins, is one of the fastest events in the animal kingdom. It is driven by the creation of very high pressure within the capsule (about 150 atmospheres!), which is in the same order of magnitude found in aluminum and steel diving cylinders (usually 220 atmospheres in a 12-liter tank). On a molecular level, the capsule's distinct wall is comprised of mini collagen fibers, designed for extremely high resistance and elasticity, that can withstand extreme mechanical pressure. The discharge process is so fast that only unique high-speed micro-cinematography devices can capture its kinetics. The discharge can be as fast as 700 nanoseconds (1 billionth of a second), creating an acceleration of up to 5,410,000 g (5.4 million times Earth's gravity!). To put this number in perspective, fighter pilots must endure up to 9 g while wearing a special g-suit that keeps blood in the upper body and prevents blackouts.[93]

The mechanisms and driving forces behind the discharge caused by a sudden increase in osmotic pressure are still unknown.[94] It is suggested that this extraordinarily high internal osmotic pressure builds at the end of capsule formation and, together with the elastically stretched capsular wall, is the main driving force of discharge. The osmotic pressure inside the capsule forces the coiled barbed thread to eject rapidly, hit the prey, and spread the toxin within it.

---

93 Alternatively, a Porsche 918 accelerates from 0–100 km in 2.6 seconds, covering a distance of 36 meters. During this same period of time, a sting cell's harpoon would cross approximately 180,000 kilometers.

94 Osmosis is the movement of solvent particles through a selective membrane, from a diluted solution into a concentrated one, in order to equalize the solute's concentration gradient.

Typical cnidarian stinging cell

The Deadly Strike of the Mantis Shrimp

Mantis shrimps are an ancient group, thought to have diverged from their closest crustaceans around 340 million years ago.[95] More than 450 species have been discovered, the majority of which are found in shallow tropical seas. All are active, aggressive predators, with a highly developed visual system and a specially adapted pair of legs (claws) located on the chest. By reference to this adapted pair of legs, mantis shrimps can be separated into two main groups. In one group are the smashing mantis shrimps, whose legs are shaped into club-like "hammers," that can strike with outstanding power. In the other group are the spearing mantis shrimps, whose legs are used as deadly, sharp spears. Both groups can unfold their legs at remarkable velocity to hit their prey.

95 Crustaceans are a large, arthropod group which includes mantis shrimps, crabs and lobsters.

Within just a few microseconds, smashing mantis shrimps can generate forces that are thousands of times their own body weight to break the shells of mollusks and crabs. To produce this extreme movement in the water, they utilize a power amplification "click" mechanism containing elastic springs, latches and lever arms.

Professor Shila Patek, from Duke University, and her colleagues used highly sophisticated videography systems to analyze the powerful strikes of the mantis shrimp. They found that the extreme speed of these strikes creates a phenomenon called cavitation. This is created by rapid pressure changes in the surrounding liquid caused by the high-velocity water jet that forms when the shrimp's claw snaps shut, creating tiny, vapor-filled bubbles. When these bubbles collapse they release a shock wave of heat, luminescence and sound, which leads to failure and flaking of surface materials. The power of this action ranges from 500 to 1,500 newtons (0.22 caliber bullet force!), delivered within microseconds. For a fraction of a second, during the collapse of the cavitation bubbles, the temperature reaches approximately 4,700°C, nearly the temperature of the sun. This deadly and effective process is achieved using a lightweight "tool" (with a mass of approximately two toothpicks) to fracture ceramic material.

Muscles alone are insufficient to provide the energy required for such a powerful strike. Thus, mantis shrimps have evolved a saddle-shaped spring that acts as a storage and power amplification system. The input of energy comes from the muscles, which gradually and elastically load the spring with the required level of energy. This is considered a different and unique preying approach compared to the cumbersome jaws and heavy hammers of other animals.

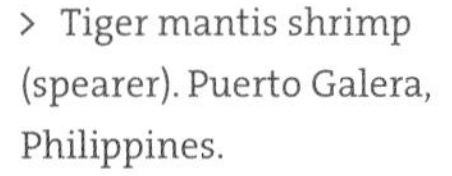

> Tiger mantis shrimp (spearer). Puerto Galera, Philippines.

^ The powerful strike of the peacock mantis shrimp (smasher shrimp) has as much force as a 0.22 caliber bullet, and for a fraction of a second, generates a temperature close to that of the sun's heat. Lembeh Strait, Indonesia.

The Occult Power of Electric Rays (Torpedo Fishes)

The term *apophenia*, coined by German neurologist Klaus Conrad (1905–1961), refers to the human tendency to see patterns or connections between unrelated things. The human brain is not satisfied with randomness. We have a profound tendency to prefer coherent stories with an overarching structure and clear endpoint. Pattern-seeking and intentionality are adaptations that have evolved as part of our sociality and the need to quickly identify risks and opportunities. This adaptation is so rooted in our minds that sometimes we overshoot, attributing hidden reality and profound meaning to coincidental events. Similarly, we tend to ascribe intentionality to inanimate objects, as reflected in the universality of animistic beliefs.[96]

This human tendency is conspicuous today in people's attraction to magicians, superpowers, comic book figures, and beliefs in telekinesis and psychokinesis.

96 Animism is a religious belief and way of relating to the world that attributes sentience and distinct spiritual essence to objects, places and creatures.

According to a 2005 Gallup survey, about 75 percent of people in the United States believe in paranormal phenomena such as extrasensory perception, telepathy and clairvoyance.[97] These beliefs are not just "folk beliefs." Many brilliant scientists sincerely believe in paranormal phenomena, including American philosopher William James (1842–1910), English mathematician and computer scientist Alan Turing (1912–1954), Swiss psychiatrist and psychoanalyst Carl Jung, and Austrian physicist and 1945 Nobel Prize winner Wolfgang Pauli (1900–1958).

One of the subsets of this deeply rooted cognitive bias is that people attribute magical powers to phenomena that occurred at a distance and have eluded traditional materialistic ways of thinking. Ancient Greeks, for example, used electric rays as a kind of anesthetic to numb pain in operations and childbirth, and believed that paranormal forces could be attributed to the fish. In fact, the Greek word for these rays is *narke*, the source of the word narcotic.[98] Roman naturalist and army commander Pliny the Elder was astonished by two types of fish with seemingly occult powers. The first was the echeneis, a legendary fish that belongs to the same suckerfish family that commonly attaches themselves to ships. Based on folk belief, this suckerfish could bring ships to a standstill "with no effort of its own." The second fish associated with occult power is the electric fish, which could numb, stun or paralyze anyone who touched or got close to it. Relying on Aristotle, Pliny and others attributed to this fish healing qualities such as relaxation of the bowels, treatment of headaches, and accelerating childbirth. Scribonius Largus (1–50 BCE), physician to the Roman emperor Claudius, described the therapeutic use of electric ray shock for easing gout and headache pains:

> *For any type of gout a live black torpedo should, when the pain begins, be placed under the feet. The patient must stand on a moist shore washed by the sea, and he should stay like this until his whole foot and leg up to the knee is numb. This takes away present pain and prevents pain from coming on if it has not already arisen. In this way Anteros, a freedman of Tiberius, was cured.*[99]

φ φ φ

97 https://news.gallup.com/poll/16915/three-four-americans-believe-paranormal.aspx (retrieved March 6, 2021).

98 http://www.elasmo-research.org/education/shark_profiles/torpediniformes.htm (retrieved March 6, 2021).

99 Largus, quoted by Piccolino, 2007.

‹ Electric rays (torpedo fishes) such as this one can produce electric discharges of up to 200 volts, paralyzing their prey. Eilat, Israel.

Physiologically, electric rays are geared with unusual prey and defense mechanisms. They can produce an electrical discharge of up to 200 volts, strong enough to knock down a human adult, and use it to paralyze prey or defend against predators. Two large, oval-shaped electric organs generate this discharge on either side of the head, where the current passes from one side of the body to the other. It is also believed that the organs that produce the electric discharge are used for sensing and navigation.

Besides the long tradition of using electric rays for therapeutic purposes, understanding the unique electric mechanism of electric rays and eels paved the way to two groundbreaking ideas during the Age of Enlightenment. The first was the understanding of the nature of nervous conduction by Italian polymath Luigi Galvani (1737–1798); the second was the invention of the battery by Italian physicist Alessandro Volta (1745–1827). In other words, electric rays are no ordinary fish; they have played an integral part in human therapeutic and scientific advancement.

Shapeshifters: The Dynamic Mimicry of the Mimic Octopus

In *The Teachings of Don Juan: A Yaqui Way of Knowledge* (Castaneda, 1968), controversial anthropologist Carlos Castaneda describes the shapeshifting of an evil person who practices black sorcery on a coyote. This description is not unique; the motif of shapeshifting, a change in the physical form or shape of a person or animal, is widespread and often appears in modern comic books, probably a relic from a time when human ancestors co-evolved with the non-human world. A universal common element in most shamanistic traditions is based on the shaman's shapeshifting to a repertoire of people, animals or ancestors' spirits. Shamans are probably highly observant of animal behaviors and inspired by insect metamorphosis and snakes that shed their skin; both of which might signify rebirth or different phases of the animal's life cycle.

φφφ

A variation of mimicry is dynamic mimicry, referring to an animal's ability to use a repertoire of defense mechanisms that mimic another species' behavior. The mimic octopus (*Thaumoctopus mimicus*) may be the ultimate example of dynamic mimicry in the form of shapeshifting. This fascinating creature was discovered as recently as 1998 off the coast of Sulawesi, Indonesia. The mimic octopus is an active moving predator but lacks a defense mechanism such as poison or a shell. So how could such a vulnerable, soft-bodied animal actively forage for food in a predator-rich environment?

The mimic octopus has developed a unique strategy that resembles the shapeshifting of the X-men comic book hero Mystique, who can mimic any person's appearance with high precision. The mimic octopus can imperfectly mimic the behavior of at least four different species, including toxic flatfish, venomous sea snakes, lionfishes and jellyfishes. The fact that all these animals have toxins suggests that the mimic octopus mimics them in order to deceive predators.

Alongside its excellent crypsis and phenotypic plasticity, the mimic octopus demonstrates an unusual, unexplained phenomenon of displaying bold and conspicuous brown and white stripes, easily drawing a potential predator's attention.

It is also active in daylight and in open sand habitat, two factors that expose it to predators. The survival of species that have adapted a highly conspicuous appearance represents an evolutionary puzzle, because it put the mimic octopus under significant preying risk. One hypothesis is that the imperfect dynamic and distinctive mimicry capabilities of toxic species are sufficient to cause potential predators to hesitate before attacking, long enough for the mimic octopus to sneak away.

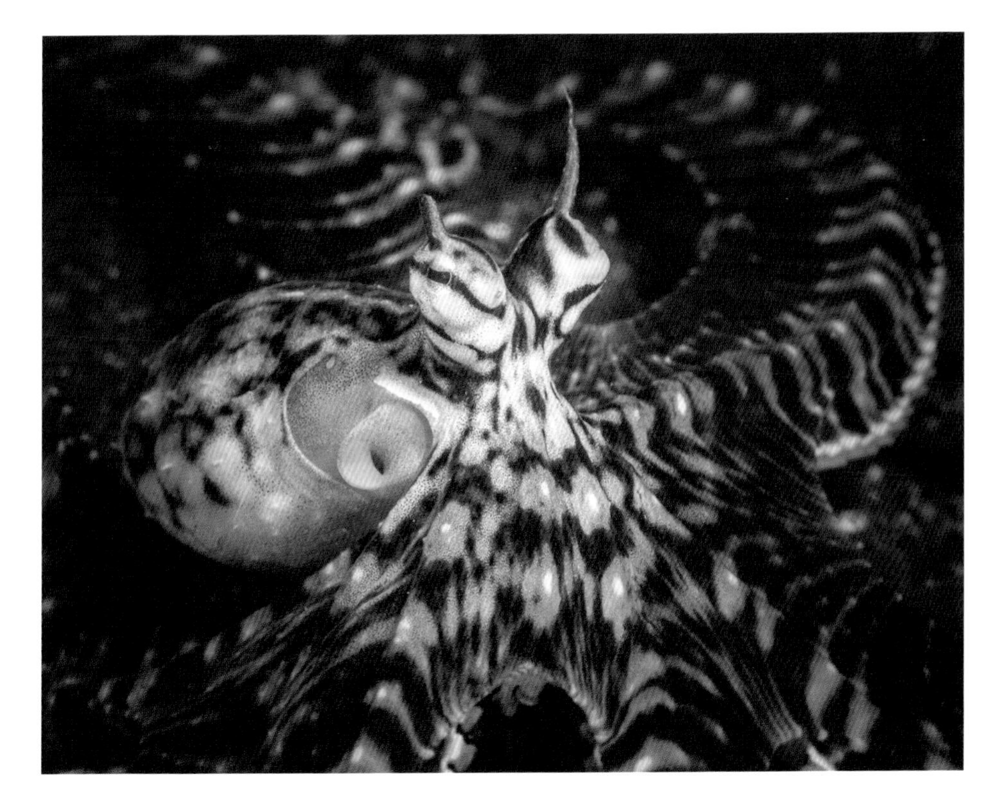

‹ The mimic octopus is capable of imperfectly mimicking the behaviors of several species, including the toxic flatfish, venomous sea snake, lionfish and jellyfish. Lembeh Strait, Indonesia.

The Aggressive Mimicry and Fast Gulp of Frogfishes

Frogfishes, a typical dweller in coral reefs, use an aggressive mimicry mechanism to attract potential prey by using a unique lure: an adapted first dorsal spine, composed of a rod and a bait. The bait is a flexible piece of tissue that resembles common food such as a worm, shrimp, or small fish. When prey gets into the "preying zone," the frogfish dangles the bait; if the prey is tempted and gets closer, the frogfish gulps it in the blink of an eye. The size and pattern of the lure are distinct for every frogfish species. The pattern of the bait is not the only factor that attracts prey. One reef dweller species, the hairy frogfish (*Antennarius striatus*), was spotted using a fluorescent worm-shaped lure. It has been reported that another species uses bioluminescent bacteria, probably for the same purpose. Interestingly, the bait can regenerate if damaged, in a process that takes 4–6 months.

Frogfishes gulp their prey using the gape-and-suck feeding mechanism in less than 6 milliseconds(!), which is faster than any other vertebrate predator. This fast gulping is achieved by the rapid expansion of the gill cavity and mouth, which creates negative pressure (suction) and inflow of water and prey into the mouth. By comparison, the stonefish (*Synanceia verrucose*), the next fastest gape-and-suck feeder, gulps in approximately 15 milliseconds. Moreover, frogfishes can enlarge their mouths by a factor of 12(!), which is significantly more than other species, and can gulp prey up to its own length by doubling the size of its stomach.

Despite its relatively slow, strange and limited locomotion, the wide variety of adaptations such as ultrafast feeding mechanisms, the ability to lure prey by using various bait patterns and illumination mechanisms, color change and perfect camouflage make the frogfish a deadly and fascinating creature.[100]

⌄ A giant frogfish flicks its lure (esca) to attract prey before gulping it down in the blink of an eye. Eilat, Israel.

100 To those of you who adore frogfishes, a comprehensive book titled *Frogfishes: Southeast Asia, Maldives, Red Sea*, written by Teresa (Zubi) Zuberbühler, can be downloaded (subject to certain terms and conditions) at: http://www.critter.ch/frogfish-book.html (retrieved March 6, 2021).

Venomous Coral Reef Creatures

Coral reefs are home to some of the planet's oldest known toxic animals, spreading across several phyla. The phylum Cnidaria, for example, is considered the oldest venomous animal lineage, around 600 million years old. Other reef animals developed various methods of avoiding predation, along with the ability to incapacitate or kill their enemies.

In general, toxic animals are divided into two major categories, poisonous and venomous. An animal is considered poisonous if it contains toxins in its body that cause harm when touched or eaten. To be hurt by a poisonous animal, one has to ingest or absorb the toxin. An animal is considered venomous if it has specialized mechanisms to deliver toxins through bites, spines or stings. For example, nudibranchs are poisonous when eaten. By contrast, stonefish are venomous, as they have spines that deliver toxins to a predator's body.

Fish venom

Fish venom systems have convergently evolved 19 times, with more than 2,900 venomous species, which account for more than half of all venomous vertebrates. The primary ecological function of using venom is either defense (the vast majority) or predation (only a few). Only in rare cases is venom used for intra- and interspecific competition, such as in the case of blennies, known for being territorial and aggressive.

The primary hypothesis behind the evolutionary source of venom glands as a defense measure in fish assumes that this development results from epidermal cells (the outermost layer of the skin) that produced antiparasitic toxins near defensive spines. At some point in the evolutionary history of fish, the venom's amount and effectivity became sufficient to deter predators.

Venom evolution is the result of antagonist coevolution. In order to cope with predators' evolutionary pressure, prey can defend itself through two major strategies, avoidance or anti-predator defense. While camouflage and crypsis are common defense strategies against detection, venom may be the optimal defense strategy in the case of a large number of predator–prey interactions. In such cases, the chemical defense is effective during the last stage of predation, following detection by the predator, but before consumption.

The energetic cost of producing toxins and its impact on survivorship might also explain the evolutionary drivers behind crypsis and aposematism. Both enable the organism to use venom only on rare occasions. Replenishment of toxins after use can take from several days to weeks, putting the fish in a vulnerable state, with possible temporarily reduced fitness. For example, in some stonefish species, full venom replenishment can take more than a month, depending on their diet.

The advantage of venom conservation is one of the factors that explains the evolution of warning coloration. Aposematic animals warn predators of their toxins in advance, thereby decreasing the frequency of their use of venom. As a matter of fact, aposematism is relatively uncommon, and changes from aposematic to cryptic lineages are significantly higher than in the opposite direction. The reason for this is that crypsis is more adaptive in the case of weak venom. On the other hand, highly potent venom contributes to preserving aposematic warning signals over the cryptic alternative, as reflected in the positive correlation between toxicity and conspicuousness.

Venom and mimicry

Venomous and poisonous creatures are strongly associated with the phenomenon of mimicry, which is the resemblance of one organism to another organism or object. The primary evolutionary goal of mimicry is to appear as either harmful or harmless (in case of aggressive mimicry) in order to deceive either predators or prey.

Müllerian mimicry is based on the principle that deterrence is achieved following frequent encounters with the threat. Warning coloration, for example, must be encountered multiple times by predators before it is associated with unpalatability. Different and sometimes unrelated species might coevolve mutualistic relationships and co-mimic each other in order to statistically increase the number of encounters with shared predators, thereby accelerating the learning curve of that predator. Müllerian mimicry is represented in the reef by certain blenny species with venomous fangs that resemble one another (for example, the striped/lined/shorthead fangblenny).

Batesian mimicry is a kind of mimicry where a harmless creature has evolved to mimic the warning signals of a harmful (usually toxic) creature, without any accompanying defense.[101] There are many examples of Batesian mimicry in the reef. One of these is the false scorpionfish (*Centrogenys vaigiensis*), also known as the prettyfin. The similarity between the false scorpionfish and the scorpionfish is so astonishing that one can only distinguish between them after careful examination of

101 Batesian mimicry is named after the nineteenth-century British naturalist Henry Walter Bates (1825–1892), who was the first to describe it.

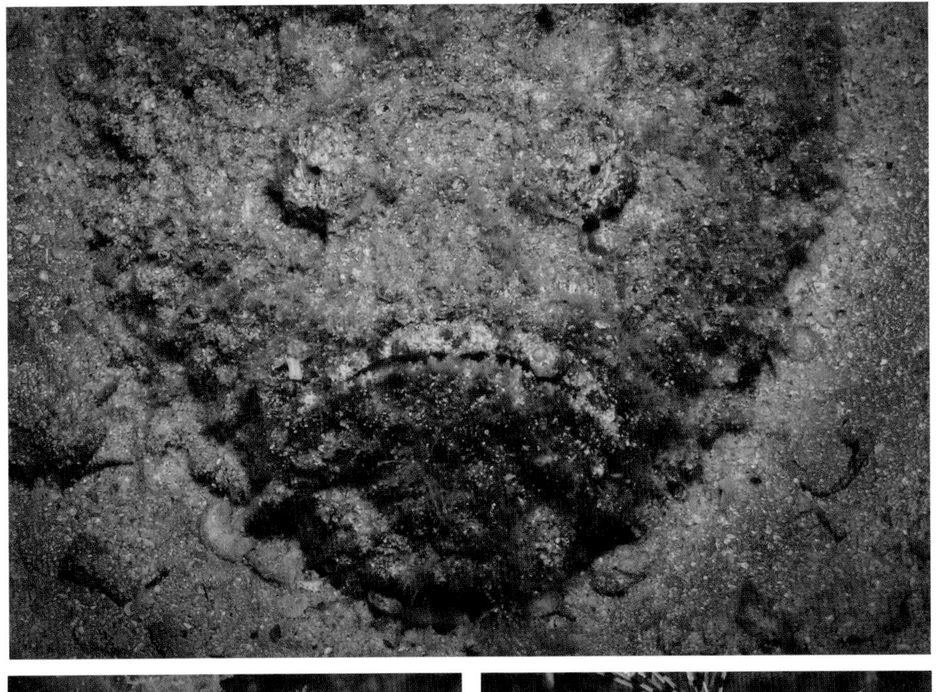

‹ The reef stonefish, found in the Indo-Pacific and Caribbean Sea coastal waters, is one of the most venomous fish in the sea. It usually waits for prey to get close to its preying zone and then strikes with incredible speed. The stonefish delivers its venom through a row of spines on its dorsal region that can be erected when it feels danger or is stepped upon. The venom is a defense mechanism that compensates for slow locomotion. Symptoms of the venom released into the human body include severe pain at the site of the puncture wound, hypotension, respiratory distress and, in extreme cases, death. Eilat, Israel.

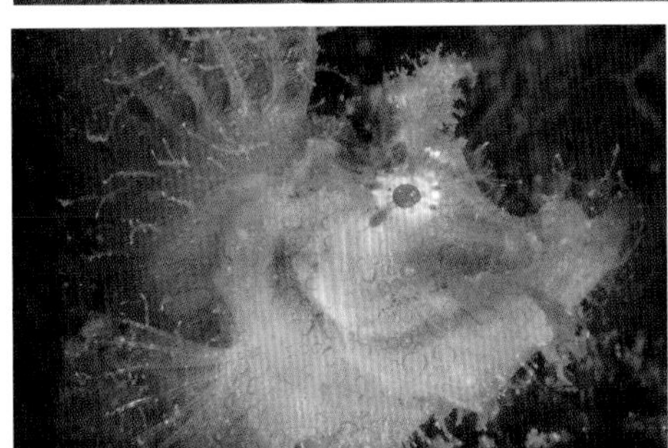

« Weedy scorpionfish. Lembeh Strait, Indonesia.

‹ Lionfish. Eilat, Israel.

‹ Scorpionfish. Lembeh Strait, Indonesia.

the fins and spines. There are also a few harmless blennies, such as the Red Sea mimic blenny (*Ecsenius gravieri*) that look and act like the venomous fang blenny.[102]

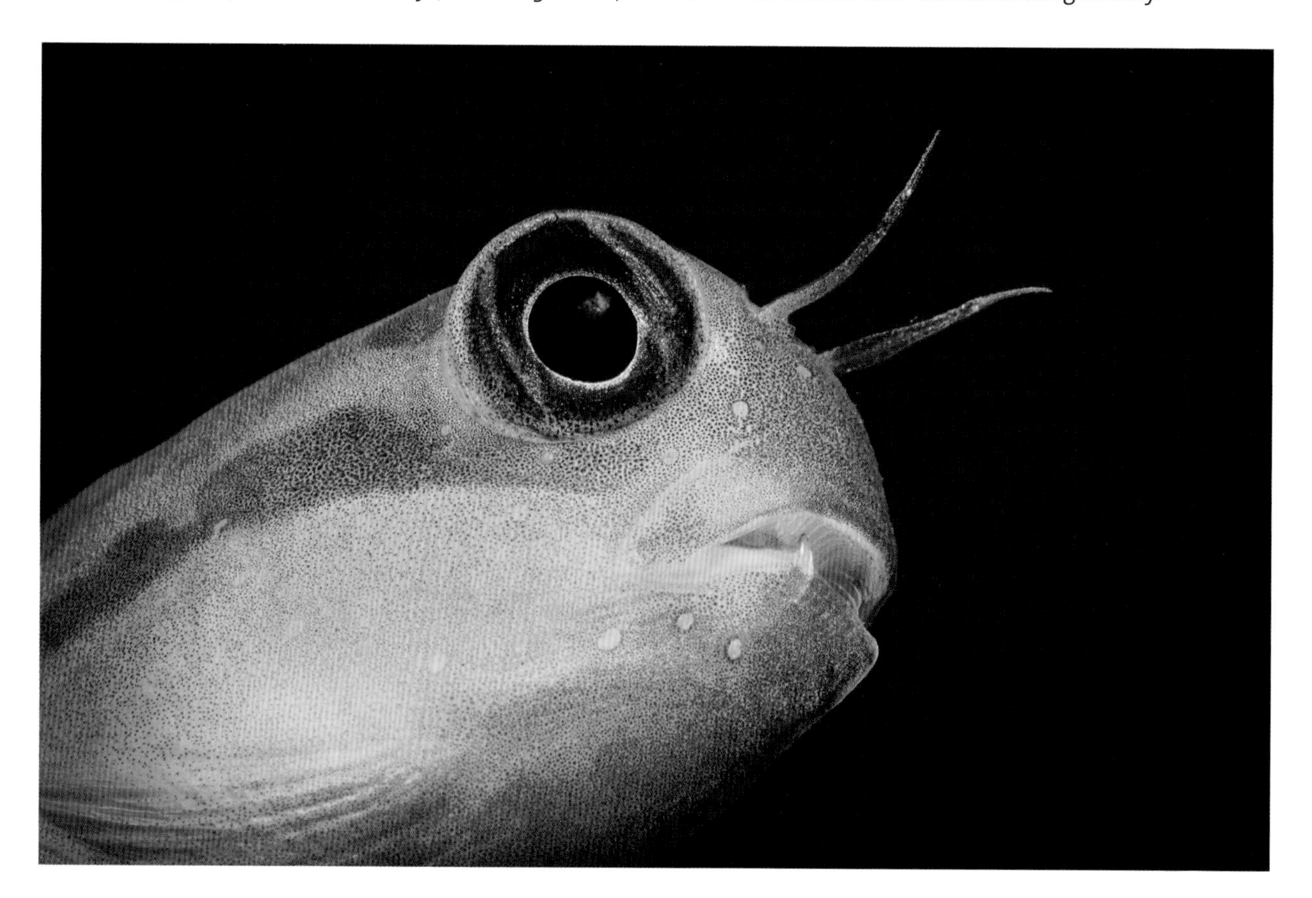

⌄ Midas blenny. Eilat, Israel.

Nudibranch poison

Many marine invertebrate species, some of them encompassing an entire phylum, are known for their toxicity. There are a few mechanisms by which marine invertebrates develop toxicity. In some cases, the toxin is produced and encapsulated by specific cells or glands. In other cases, invertebrates can accumulate toxins, in either their guts or skin, from a diet that includes microbial toxins, or by using natural toxic products from other invertebrate species upon which they feed.

Nudibranchs are incredibly vulnerable to predators since they shed their natural protection, a hardened shell, after the larval stage. They serve as prey to a wide variety of marine animals, including other nudibranchs. Hence, nudibranchs have developed

102 https://www.tfhmagazine.com/articles/saltwater/mimicry-in-reef-fish-communities-full (retrieved March 6, 2021).

other defense mechanisms to deter potential predators. Some of them rely on camouflage or aposematic coloration and the production or ingestion of toxic material.

A subgroup of nudibranchs called Aeolids (for example, *Flabelina*) usually feed on cnidarians members and store their stinging cells; when predators try to touch them, these stinging cells are "fired" at the predators to deter them. The question of how nudibranchs avoid triggering these cells when they are being ingested remains unanswered.

Another common nudibranch group, *Chromodoris*, includes colorful aposematic nudibranchs that digest several toxic compounds from sponges and then redistribute and store them in their mantle rim, a place where predators frequently target their attack. As with the Aeolids mentioned above, the mechanism by which they ingest such toxic materials without causing internal damage is puzzling. It is suggested that the resemblance of the color and pattern of many species of *Chromodoris* nudibranchs, as well as the presence of the same potent compound, is an example of Müllerian mimicry.

˅ The red-lined flabellina aeolid nudibranch acquires its defensive weaponry by ingesting cnidarians and storing their stinging cells in the tip of their finger-like cerata. Lembeh Strait, Indonesia.

>> The magnificent chromodoris ingests toxins from sponges. Eilat, Israel.

> Unidentified dorid nudibranch with conspicuous colors. Puerto Galera, Philippines.

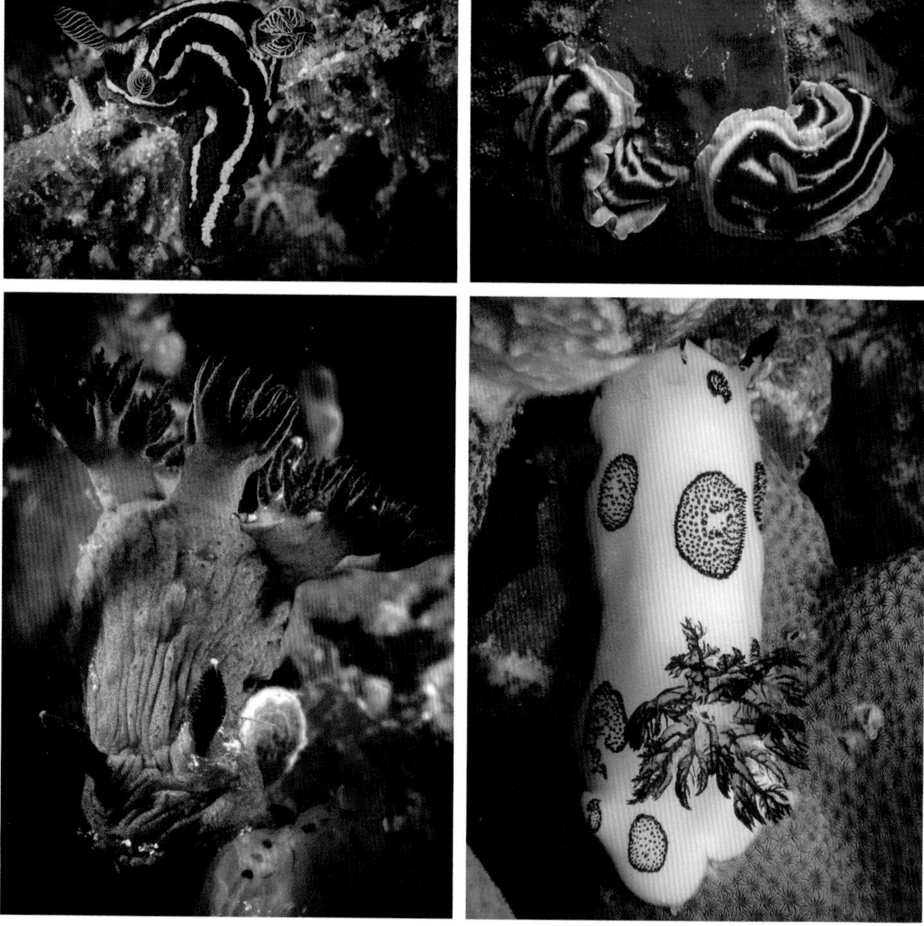

>> Funeral jorunna nudibranch. Puerto Galera, Philippines.

> Unidentified genus nembrotha nudibranch. Puerto Galera, Philippines.

> Most nudibranchs are predators that feed on sessile animals such as sponges or corals using a tongue-like organ called a radula to rasp and grasp. Unlike other nudibranchs, the green melibe (*Melibe viridis*) has an exceptional mechanism that actively preys on moving animals such as small crustaceans and plankton. The melibe has a large extensible bubble-like oral hood that scans the ground and captures prey. When the prey is inside the hood, the melibe's tentacles lock it in and the hood compresses, pushing water out until the prey is forced into the melibe's mouth. Lembeh Strait, Indonesia.

Zombie Powder: TTX Venom Mystery

In 1772, James Cook was commissioned to lead his second voyage to search for the hypothetical continent Terra Australis. In his 1777 book, *A Voyage Towards the South Pole and Around the World*, Cook describes a terrible experience that he had after eating a certain type of fish purchased by his clerk from one of the locals:

> *It was of a new species, something like a sun-fish, with a large long ugly head. Having no suspicion of its being of a poisonous nature, we ordered it to be dressed for supper; but, very luckily, the operation of drawing and describing took up so much time, that it was too late, so that only the liver and row were dressed, of which the two Mr Forsters and myself did but taste. About three o'clock in the morning, we found ourselves seized with an extraordinary weakness and numbness all over our limbs. I had almost lost the sense of feeling; nor could I distinguish between light and heavy bodies, of such as I had strength to move; a quart-pot, full of water, and a feather, being the same in my hand. We each of us took an emetic, and after that a sweat, which gave us much relief. In the morning, one of the pigs, which had eaten the entrails, was found dead. When the natives came on board and saw the fish hanging up, they immediately gave us to understand it was not wholesome food, and expressed the utmost abhorrence of it; though no one was observed to do this when the fish was to be sold, or even after it was purchased.*

Captain Cook was fortunate; the fish he and his crew member ate was probably a species of pufferfish and his experience is one of the earliest written descriptions of the effect of tetrodotoxin (TTX), a potent neurotoxin that is widely distributed in marine animals and in some terrestrial (mainly amphibian) species. Approximately 140 animal species contain TTX, including the blue-ringed octopus, pufferfish (*fugu* in Japanese), Central American harlequin frogs, sea stars, several snails, flatworms and more.

TTX can cause severe harm to humans. The blue-ringed octopus, for example, carries enough poison to kill several people within minutes. The toxin's chemical mechanism blocks voltage-gated sodium ion channels, preventing the entry of sodium ions into the nerve and muscle tissues. Prevention of sodium ions from entering the cells causes the electrical signaling in the nerves to shut down, leading to severe physiological consequences.

Until recently, it was a mystery how such diverse, unrelated terrestrial and marine organisms could all evolve the same toxin. One suggested reason was convergent evolution, which is an independent evolutionary trajectory of similar features in species that creates analogous structures and characteristics. However, it was recently discovered that the presence of TTX in most of these animals is because they all consume or live in symbiosis with bacteria that produce it.

φ φ φ

In some cultures, the effects of TTX have been known for over 5,000 years, mainly due to human consumption of species with TTX. Early Egyptians, for instance, were aware of the risk of eating pufferfish, and Chinese herbal medicine refers to the merits and risks associated with the flesh and eggs of these fish.

Fugu, a Japanese delicacy made with pufferfish, has special cultural meaning. It is one of the most expensive delicacies in Japan, and a single multi-course meal can cost hundreds of dollars. If not prepared correctly, it can cause severe illness or even death. A *fugu* chef must go through intensive training before being granted the license to prepare *fugu* for others. An unconfirmed myth says that if a *fugu* chef mistakenly kills a customer, he is honor-bound to commit suicide in a ritual known as *seppuku*,[103] using a *fugu* cutting knife.

Another intriguing aspect of TTX is its use in Haitian voodoo culture as a zombie drug. There is an unusual story about a Haitian man named Clairvius Narcisse who admitted himself to an American hospital in Haiti in 1962. Narcisse suffered from fever and other severe symptoms; after three days, he was pronounced dead and the next day, he was buried. Eighteen years later, his sister came across a man in her village marketplace claiming to be Clairvius Narcisse. The man said that shortly before he had been pronounced dead, he was enchanted and paralyzed by a sorcerer. He claimed that he had been lucid during the entire burial ceremony, and although he had been without will, unable to move or speak,[104] he did hear his sister weeping at his death. Narcisse said that he was later stolen from the tomb by a *boke* (sorcerer) and forced to work with other zombies as a slave in a sugar plantation. After two years, the *boke* was murdered by one of the zombies, which enabled Narcisse and the other zombies to escape. Clairvius spent the next 16 years wandering the Haitian countryside before meeting his sister.

---

103 A ritual performed by self-disembowelment, sometimes known as hara-kiri.

104 Based on Haitian belief, the stolen spirit is kept by the *boke* in a jar or bottle and the victim is left without the capacity to act independently or make decisions.

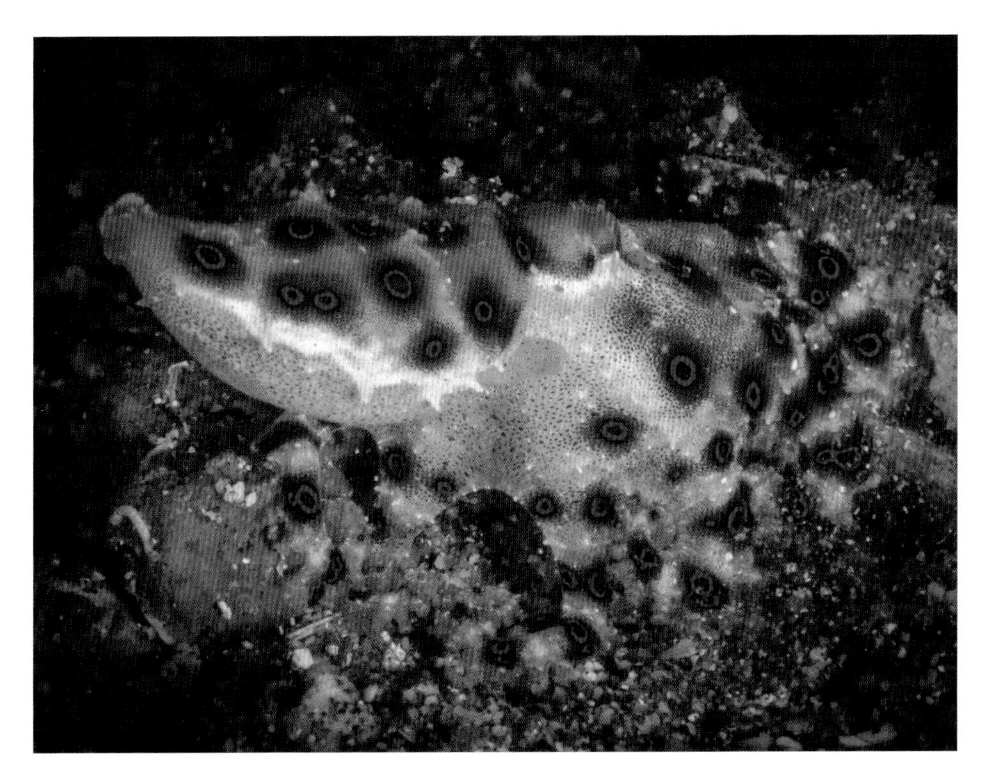

‹ The beautiful blue-ringed octopus is among the deadliest animals in the sea. Its salivary glands host dense colonies of TTX-producing bacteria, providing the bacteria with ideal living conditions while using the toxin they produce for defense and preying purposes. Lembeh Strait, Indonesia.

‹ Black-spotted pufferfish. Dauin, Philippines.

Between 1982 and 1984, a young Harvard ethnobiologist named Wade Davis traveled to Haiti in search of the secret of zombie poison. Davis interviewed a few *bokes* and collected samples of zombie powders. He found that these powders included TTX ingredients extracted from pufferfish and suggested that the symptoms of TTX poisoning are very similar to what Narcisse had exhibited when hospitalized; it induced a coma that mimicked the appearance of death. Moreover, Davis argued that there are similarities between the symptomatology of Haitian zombies and *fugu* poisoning victims.

The hypothesis that Haitian zombies are made through the use of TTX present in pufferfish was criticized by the scientific community, who argued that it was "bad science" that led Davis and others to rush to conclusions regarding the effects and traces of TTX as a zombie drug. Nevertheless, it seems that the question of zombification in Haiti should be left open, for many reasons, including the widespread Haitian beliefs in zombies, the appearance of TTX in some zombie powders that have been analyzed, the known effects of TTX to induce suspended animation that is almost the same as death, and the well-known effect of predisposed culturally induced drug experiences.

Crowd Behavior: Why Do Fish School?

Elias Canetti (1905–1994), winner of the 1981 Nobel Prize for Literature, was enthusiastic about studying the power of psychopathology as manifested by crowd behavior and mass movements. In his seminal book *Crowds and Power* (1962), Canetti managed to clearly describe the power of the crowd and the variety of ways in which people are subordinate to it. He believed that the most striking phenomenon about collective movement such as mass flight is "the force of its direction," which derives its power from its coherence. Canetti captured the tension between human individuality and the attractiveness of being part of a crowd. In his opinion, crowds discharge humans' primordial fear of the unknown and alleviate the instinct to avoid contact with strangers. Because of the effect of unique cultural and human cognitive perception, the collective movement of humans is not identical to that of animals (including schooling fish), but it does share similarities. For example, with humans as with schooling fish, an organism senses its world as part of a whole, part of an organized and unified species, a characteristic that resembles a superorganism.

φφφ

Fish schooling is a common phenomenon. More than 25 percent of fish species form schools at some point in their life. The evolutionary transition to schooling provides significant adaptive advantages in foraging, reproduction and migration. For decades, scientists have tried to explain two central questions regarding fish schooling. The first relates to how a school is formed; the second refers to its adaptive benefits.

The first question can be partially answered by mathematical models that simulated interactions between individuals that compose a group. One of these models is based on three simple rules of engagement that apply to humans as well as fishes:

I. Attraction – Move toward your neighbors in a remote zone.
II. Repulsion – Move away from your neighbors in a nearby zone.
III. Alignment – Match the speed and direction of neighbors in an intermediate zone.

As a side note, pioneering Hollywood moviemakers adopted a similar model and applied it in animated movies such as *The Lion King* and *Batman*.

This model also shows that schooling is formed by self-organization that does not require the presence of leadership. This characteristic can explain human and animal collective movement without being subject to a priori cognitive action or neural structures. An example of this self-organization character is the formation of a school from a shoal of foraging fishes that is threatened by a predator.[105] Once some of the fishes identify the threat, they swim in a similar direction, and in an example of positive feedback, gradually recruit more and more fishes that eventually form a coherent, polarized moving school.

The second question, one that still needs to be answered, is what are the advantages of schooling? This question is addressed mainly at the group rather than the individual level, from an evolutionary perspective.

In general and for each fish, the chance of finding itself in the mouth of a predator decreases as the size of the school increases, due to a number of advantages. The first advantage is that, in both schools (polarized) and shoals (non-polarized), the potential predator might be confused by the large size of the group. The second advantage of schooling is the 360 degree vision, which increases the chance of detecting a potential threat and the efficacy of foraging for food. The third advantage is that

105 Unlike schooling, where fish swim in the same direction (polarization), shoaling means fishes swim randomly and in a non-coordinated direction.

> School of venomous striped catfish. Puerto Galera, Philippines.

v School of striped eel catfish. From a distance, it is difficult to distinguish between the venomous striped eel catfish and the harmless juvenile convict blennies that mimic them. Bunaken, Indonesia.

schooling improves swimming efficiency because the fishes in front form an eddy that minimizes water resistance for the fishes behind (similar to a cyclists' peloton, which significantly reduces air resistance). The fourth advantage of schooling is its effect on the sensory system of potential predators. Coordinated and synchronized fish movements emit sound and pressure waves that may confuse predators' sensing systems, thereby disrupting the senses generated in the inner ear and lateral line organ.[106] Moreover, prey fishes that move close to each other might blur predators' electro-sensory perception. Examples of marine predators with such organs can be found in sharks and rays; both have electro-sensory organs called ampullae of Lorenzini, which helps them sense electric fields in the water that can potentially be disrupted by schools of fish.

It is important to mention that fish schooling comes with disadvantages as well; for example, some marine predators have developed expertise in hunting for schooling fishes. Moreover, the interactions between behaviors and the group's emergent properties are still poorly understood. In some instances, there is a conflict between the fitness level of the collective (group selection) and that of the individual.

< School of sea goldie (scalefin anthias) around yellow scroll coral. Sinai Peninsula, Egypt.

106 The lateral line organ is a row of sunken sensors, running from the gills to the tail, present in all aquatic invertebrates, including fishes. This organ is responsible for detecting movement, vibration and environmental pressure changes.

"Without contraries is no progression. Attraction and Repulsion, Reason and Energy, Love and Hate, are necessary to Human existence."

William Blake, *The Marriage of Heaven and Hell*, 1790

CHAPTER 7

Reproduction
The Wild Side of Sex in Coral Reefs

« A pair of mating mandarinfish.
Puerto Galera, Philippines.

In the summer of 2013, I decided to travel to the north part of Sulawesi, Indonesia. My final destination was Lembeh Strait, the mecca of muck diving, and one of the world's best locations for macro photography. Before I arrived at the diving resort, and as part of my interest in the anthropology of religion, I decided to explore the unique burial rituals of the Turaja, an indigenous, south Sulawesi ethnic group. The Turaja live about 12 hours away from the Makassar airport, so I decided to visit the area I would pass through on my way there. It is mostly inhabited by the Bugis people, known for their traditional skills as shipbuilders, pirates and slave traders. The Bugis settled in Sulawesi about 2,500 years ago, part of a sizeable Austronesian migration from southeast China. Today, most Bugis identify as Sunni Muslims; however, their religious practice, influenced by Sufi tradition, is a syncretic blend of Islam and Austronesian shamanistic culture that involves the giving of offerings to local spirits and deities.

The Buginese language distinguishes between five types of gender (how one expresses sexual identity in a cultural context) and sexuality (one's sexual interest), namely: "female-women," "male-men," "female-men," "male-women," and "transgender priests," the latter known as a *bissu*. *Bissus* serve as clerics and healers and are considered saints.

Bissus are androgynous, imagined as embodying both female and male characteristics, and dress in a manner that emphasizes both male and female aspects. Historically, *bissus* were in charge of performing sacred royal rituals, communicating with ancestors' spirits, and safeguarding sacred texts. *Bissus* also act as shamans and have connections with the spirit world, usually via an altered state of mind such as a trance. They perform blessings for the local communities at almost every event, including rice planting, before harvest, and at marriages. Ironically, they also bless people before they go on Hajj, the Islamic pilgrimage to Mecca. In order to give blessings, *bissus* often go into a trance, inviting the appropriate deity to enter their body by performing an elaborate ritual that includes dancing, chanting and food offering.

> A small number of transvestite *Bissu* priests are an integral part of the Bugis community, performing rituals directed to local deities. Sulawesi, Indonesia.

The tolerance of this predominantly Muslim society towards non-conventional sexuality intrigued me, as it was not in line with the gender taxonomy of modern Western culture. It is worth mentioning that in many ancient cultures, including ancient Greece (pederasty),[107] Persia, the Ottoman Empire,[108] and historical and contemporary Asian cultures,[109] diverse sexualities were culturally acceptable; they did not occupy a central role in people's social and personal identity, and were considered mere sexual preferences. North American cultures were no exception and many Native American peoples included third gender people ("two-spirit people") who undertook significant cultural roles. For most of history, many societies did not perceive human sexuality in fixed and unidimensional terms. In this regard, French historian and philosopher Michel Foucault argued that sexuality, as a prime essence of selfhood, emerged only with modernity and Western-influenced heteronormativity during the late-nineteenth century.

107 In ancient Greece, pederasty referred to the romantic relationship between an adult man and a boy.

108 In Ottoman culture, boys were considered to be a distinct gender.

109 Such as the *Hijra* in India, which is both a religious sect and a caste.

The tolerant approach described above emphasizes the central role of social discourse and society in sexual and gender identity throughout history. For most people, androgyny is perceived as non-conventional sexuality; by comparison, marine animals encompass a remarkably wide variety of sexual reproduction strategies that are far more diverse than we see in humans. The occurrence of two clearly separate sexes, male and female, is not mandatory in the animal world. Moreover, the idea that sex is defined by typical physical male and female characteristics (phenotypic sex) is not relevant in the animal world. Some animals change their sex during their lifetime; some animals are both female and male at the same time. In other words, when it comes to marine sex, there is no such thing as "normal." Sexuality takes diverse forms, going far beyond the traditional Western Mars–Venus dualism. The only thing that matters in nature is employing a reproduction strategy that maximizes each species' odds of transferring their genes to future generations.

φ φ φ

Of all biological processes, reproduction is most fundamental. Bacteria, the simplest and most primordial living cells on Earth, do not reproduce by sex. They reproduce by splitting into two, a form of asexual reproduction.[110] On the face of it, it would make sense to infer that asexual reproduction would spread rapidly through a population, meaning sexual reproduction would eventually become extinct, because it requires far less energy than sexual reproduction. For example, in an asexual species, organisms can rapidly reproduce independently, without the need to look for a mating partner.

Nevertheless, sexual reproduction is widespread, mainly because it creates and maintains a high degree of genetic diversity, which in turn creates an increased number of phenotypic variations upon which natural selection acts. By filtering the advantageous genotypes, the process of natural selection is accelerated. In other words, sexual reproduction is costly, but also appears to be essential for the long-term survival and adaptation of complex creatures.

Among animals, almost all species practice sexual reproduction; some, such as cnidarians, can reproduce both sexually and asexually. However, to overcome the potential drawbacks of costly sexual reproduction in the sea, many marine creatures

110 In asexual reproduction, offspring originate from a single organism, inheriting only the genes of that organism. In sexual reproduction, the fusion of reproductive cells from two organisms creates offspring with diversified genetic material from both parents.

have evolved various sexual reproduction strategies. These include sex change, hermaphroditism, male brooding, sperm depositing, broadcast-spawning, and more.

> Female mantis shrimp brooding fertilized eggs. The black dots on the eggs are the developing eyes of larval shrimps. Lembeh Strait, Indonesia.

Hermaphroditism: Sex Change in Fishes

Hermaphroditism is named after the son of the Greek god Hermes and the goddess Aphrodite, the gods of male and female sexuality. Various legends exist of the god Hermaphroditus, with the common theme describing him as a union of a female body with a male sex organ. According to one of the legends, the nymph Salmacis was charmed by the handsome young Hermaphroditus and prayed to the gods that they fulfill her wish to unite with him forever. Salmacis's prayers were answered and the result was the creation of a being who was half woman and half man.

Hermaphroditism plays a significant role in all sorts of mystical beliefs, and is deeply rooted in human thought. It universally signifies the union of opposites and the essence of wholeness, as well as the human pre-conscious condition before it is differentiated into male and female. It is not surprising that hermaphroditism is so central to human culture since humans tend to think in pairs of dialectic notion structures. French anthropologist Claude Lévi-Strauss (1908–2009) coined the term binary opposition structures for this universal structure of myth and human thought;

for example, darkness–light, land–sea, culture–nature, man–woman, and good–bad. Based on this structuralist approach, the meaning is not absolute but relative and depends on binary oppositions. In this regard, hermaphroditism is a term that reconciles the tension embodied in the opposite terms of male and female.

φ φ φ

Hermaphroditism, or the expression of both male and female sex within an individual's lifetime, intrigues modern biologists, both because the phenomenon is widespread among plants and animals, and also due to its evolutionary implications. For example, hermaphroditism is found in approximately 2 percent of all teleost fish species.[111] The alternative to hermaphroditism is "sex determination," characterized by sexual fate initiated during embryonic development.

Hermaphroditism is used today as an umbrella term that encompasses various sex change processes throughout an animal's life cycle. Some fish species exhibit sequential hermaphroditism; that is, an individual is born as one sex and later transforms into the opposite sex. Other species have both male and female sex organs at the same time, known as simultaneous hermaphroditism; an example of this is the hamlet fish, found in the Caribbean.

In some cases, the change in sex results from environmental influences such as temperature, pH, salinity, and social factors such as the absence of a male in the group. In other cases, sequential hermaphroditism is characterized by the change of sex during adulthood.

Change in sex can occur in either direction; from male to female or from female to male, but the latter is more common.[112] In anemonefish, for instance, a single male often lives with a single female, or with a few non-reproductive males, and the remaining fishes are immature. If the female dies, the male takes her place and becomes a female, and one of the juveniles, or another male anemonefish, takes its place.

Examples in the opposite direction, from female to male, can be found in some species of anthias, parrotfishes and wrasses that form harem groups, a group composed of several females and one or two dominant males. Harem groups are found in schools of sea goldies, usually consisting of one male, many females, and their offspring. All sea goldies are born as female. In the absence of a male, the largest female will transform into a male, a process that takes several weeks.

111 Teleosts are the largest group of fishes, including more than 95 percent of all of the present-day fishes.

112 Transformations of male to female or female to male are known as protandry and protogyny, respectively.

The transformation includes a gradual change in color, the adoption of aggressive behavior towards other males (usually from outside the harem), and the ability to mate with the females and reproduce.

One of the most compelling explanations for sequential sex change is the size advantage hypothesis. This theoretical model predicts that sex change is favored when an individual's reproductive success depends on its size or age; for instance, large females can lay more eggs than small ones; large males can breed with more females than small males. For example, in anemonefish, as described above, a larger female can produce more eggs, and a larger male has no advantage. This is why males become female when the female dies. The sequence of hermaphroditism is determined by the pace of increasing fertility with body size. In cases where the fertility of an average male increases sharply with body size, and more so than that of an average female, the animal will go from female to male.

In most cases of sequential hermaphroditism, this change only takes place once in a lifetime. However, some species are characterized by serial bidirectional sex change, which means that sex change occurs several times throughout their lifetime. An example of this can be found in some species of coral gobies, groupers and angelfish.

The mechanism of sex change is still unclear; however, it was found that in blue-headed wrasses, the change takes place in an epigenetic process.[113] The removal of a dominant male from the group induces stress, and a sex change process begins in the largest female, a process that might take less than two weeks. The social and environmental change represses the expression of genes responsible for converting male hormones (androgen) to female hormones (estrogen). As a result, there is an increase in androgen, which is responsible for the regulation, development, and maintenance of male characteristics. This is a striking example of epigenetic inheritance and phenotypic plasticity that has gained the attention of the scientific community over the last few decades, and might also occur in other species.

The main advantage of hermaphroditism is the assurance of finding a reproductive partner. This advantage is significant in the case of simultaneous hermaphroditism compared to sequential hermaphroditism. However, the drawback is that the energetic cost associated with maintaining two reproductive systems is high.

In some cases, an evolutionary driver for the hermaphroditic breeding system can be correlated with low mobility since hermaphroditism increases the number of encounters with potential mates. Thus, low mobility might be one of the evolutionary

113 Epigenetics is the process of changes in the phenotype that are not a result of alterations in the DNA sequence.

factors behind the simultaneous hermaphroditism of nudibranchs and worms. For them, encountering any member of the species means encountering a potential mating partner, regardless of its gender.

< A male sea goldie (*Pseudanthias squamipinnis*). Schools of sea goldies usually form a harem and demonstrate female to male hermaphroditism. Lembeh Strait, Indonesia.

< Female sea goldies (scalefin anthias). Lembeh Strait, Indonesia.

> It has been proposed that the lemon coralgoby (*Gobiodon citrinus*) demonstrates serial bidirectional sex change. Eilat, Israel.

Hybrids: Interspecies Breeding

The chimera is a monstrous hybrid creature from Greek mythology, that was born with a triple animal body. In the *Iliad*, Homer described the chimera as "an invincible inhuman monster, but divine in origin. Its front part was a lion, its rear a snake's tail, and in between a goat. She breathed deadly rage in searing fire."[114] For thousands of years, hybrid creatures were universal themes in mythologies throughout the world, and in many cases, superior powers were attributed to them. Hybrids between humans and animals are also ubiquitous, and well-known examples in Greek mythology include the Siren, a bird-woman, the Centaur, a human-horse, and the Minotaur, who is half bull, half human.

During the twentieth century, as part of anthropological theory development, two central areas of interest relate to hybrids. The first discusses the role of cultural categories. A significant contribution to this area was made by British anthropologist

114 Homer (2007), *The Iliad*. Translated by Johnston, I., Richer Resources Publications, Arlington, Virginia.

Mary Douglas (1921–2007). In her book *Purity and Danger: An Analysis of the Concepts of Pollution and Taboo* (1966), Douglas showed that categories play an essential role in every culture, establishing clear and stable cultural taxonomies. She highlighted the concept of pollution as representing something outside the accepted social categories. In this context, hybrids between animals and humans are usually referred to as ambiguous, unnatural and impure, representing something that challenges social classification.

The second anthropological concept, proposed by British anthropologist Victor Turner (1920–1983), is liminality. Liminality describes a threshold state which exists between distinct domains or categories. Examples of liminal processes are initiation rituals that mark the passage from one stage of life to another. One example of this is a *bar mitzva*, the Jewish coming of age ritual. In other cases, human–animal hybrids represented a reconciliation between cultural dichotomies such as order/chaos and familiar/foreign. In this respect, biological hybrids, the production of offspring by two different species, can be seen as the natural model of liminal states. Therefore, they are used as a symbol in transformation and initiation rituals and myths.

It seems that the anthropological interpretation of hybrid symbols corresponds with the biological paradigm of species definition. After all, human beings are classifying animals, and based on the contemporary postmodern view, biological and anthropological categories are socially constructed rather than reflective of objective reality.

φ φ φ

A longstanding debate in biology deals with the validity and definition of species' boundaries and raises the question of what constitutes a species. Nearly 300 years ago, Swedish botanist and zoologist Carl Linnaeus (1707–1778), the father of modern taxonomy, developed a hierarchical classification system in which species are the fundamental unit of analysis upon which all branches of biology are based. Linnaeus, who lived a century before Darwin, believed that God created 4,000 species of animals; today, we know there are millions. His classification system includes, above the species level, assemblages of related organisms that are hierarchically categorized into genus, family, order, phylum and kingdom.

Since Linnaeus, species have been the fundamental category of biological organization. The centrality of species' taxonomy to biology was echoed by notable biologist Edward Wilson, who wrote in his 1999 book *Diversity of Life*: "Not to have a natural unit such as the species would be to abandon a large part of biology into free fall, all the way from the ecosystem down to the organism."

In 1942, biologist Ernst Mayr (1904–2005) proposed the following definition: "Species are groups of actually or potentially interbreeding natural populations, which are reproductively isolated from other such groups." Mayr considered animal hybrids to be rare, and hybridization to be a negative selective agent.

In a nutshell and in most cases, speciation happens when populations evolve to become distinct species with unique characteristics. Species' formation mostly occurs due to species' development within an isolated geography, or in the same geography but in a different ecological niche. The new species eventually becomes genetically distinct, whether by genetic drift, accumulation of mutations that prevent the populations from mating, or different selective pressures. Therefore, the new species no longer recognizes the parent species as a potential mating partner, or mating opportunities become very limited due to different habitats.

Mayr's elegant definition encompasses the lion's share of animal life, and is still considered the most useful definition of species. However, in the last few decades, with the enormous increase in biological observations and advanced genetic analysis, many deviations from Mayr's concise definition have been discovered. For example, some species apparently can interbreed, sometimes even producing viable and fertile offspring. Moreover, Mayr's definition did not account for asexual reproduction, which is relatively common among some marine animals (such as cnidarians). So, similar to other so-called biological laws, exceptions can be found in Mayr's reproductive isolation criteria.

Biological hybridization is widely prevalent in plants, but often perceived as unnatural and rare in animals. Examples of hybrids in animals are mules (the offspring of a female horse and male donkey) and ligers (the offspring of a male lion and female tiger). However, meta-analysis across various plants and animals indicates that the phenomenon is more widespread than previously thought, with at least 25 percent of plant species and 10 percent of animal species hybridizing with other species. It is also worth mentioning that the genome of some European and East Asian peoples contains genetic material from long-extinct Neanderthals, indicating that hybridization was probably part of the evolution of *Homo sapiens*.

For many years, hybridization of animals in coral reefs was considered rarer than that of animals in terrestrial habitats. This assumption was due to the presence of attenuated physical barriers in marine environments, which blurred the role of biogeographic reproductive isolation as a driver of speciation. However, over the last 25 years, an increasing body of evidence shows that hybridization in coral reef habitats is more common across fishes and reef-building corals than previously thought. I describe here only two examples of hybrids among coral reef dwellers

< My son, Or, near a branching *Acropora* coral. Eilat, Israel.

(butterflyfish and *Acropora* corals); however, it makes sense to assume that there are more.[115]

In some families of angelfish (*Pomachanthidae*) and butterflyfish (*Chaetodontidae*), it has been reported that the occurrence of hybridization between species from the same family involves close to 50 percent and 40 percent of the species, respectively. The offspring are sometimes beautifully colored, with intermediate coloration between contributing parents. This hybridization usually occurs between closely related species that inhabit the same geographic area. Furthermore, a study found that butterflyfish hybrids have similar fitness and fecundity as the parent species. These findings, which demonstrate a deviation from Mayr's definition of species, might indicate hybrids' potential to serve as a source of evolutionary novelty that might increase the organisms' fitness and lead to the formation of new species.

Hybridization can also be found in corals. Many coral species reproduce sexually through broadcast spawning, the simultaneous release of eggs and sperm into the water, followed by fertilization and larval development in the open sea (see Chapter 2). In some cases (or in some localities), a few or even many species may spawn their reproductive material all at once, which may lead to fertilization between an egg and sperm of different species and the development of a hybrid.

115 There have also been observations of mating between different species of sea slugs, but this has not been scientifically substantiated (http://www.seaslugforum.net/showall/abmating (retrieved March 5, 2021)).

As with angelfishes and butterflyfishes, coral hybrids mainly occur between closely related species from the same genus. These hybrids might result from mass spawning, where the offspring may be normal-looking corals that grow and multiply via asexual reproduction and live for many years. Examples for such hybrids mostly include crosses between branching coral species of the common reef-building genus *Acropora*. In short, hybridization in corals may blur species boundaries, increasing reef diversity without speciation. In such cases, reef diversity may be enhanced, since even infertile hybrids can reproduce asexually (via budding) and form colonies that live for hundreds of years.[116] Moreover, these colonies have unique genotypes and morphologies that might enable them to cope better with selection pressure.

The consequences of hybridization on reef ecology are challenging to assess. On the one hand, hybrids may contribute to species extinction (reverse speciation) if they are unfit and sterile. On the other hand, hybrids may increase biodiversity by producing unique genotypes with new adaptive traits (new lineages) that can better cope with environmental changes and natural selection pressures.

In sum, it seems that the traditional representation of evolution as a mere "tree of life" where animals have diverged from each other, creating more and more branches that never cross, does not actually provide an accurate picture of nature. Professor Michael Arnold, in his book *Evolution Through Genetic Exchange*, suggests that "web of life" may be a better description of evolutionary diversification. In any event, it seems that hybridization might be a more prevalent mechanism in nature and a source of evolutionary novelty that provides new adaptive opportunities. A more balanced view of evolution would be a combination of the two perspectives; a tangled web, where diverging and recombining species interweave. In my opinion, the "tree of life" metaphor is still more common in evolution than the web-like interconnection between species, since without it we would not have differentiated species at all. This new approach, however, of highlighting the horizontal connections between species, is consistent with the current zeitgeist that sees nature as a web of interacting and interrelating parts (elaborated further in Chapter 9).

116 The longevity of corals can reach hundreds or even thousands of years. It has been found that in deep water (300 m and below), black corals can live for thousands of years. This makes them the oldest marine animals ever found.

A Death Sentence: Octopus Sex

In Greek mythology, Sirens are half-bird, half-woman creatures who sing enchanting music to lure sailors to their deaths. As wonderfully described by Homer in the *Odyssey*:

> *First you will raise the island of the Sirens, those creatures who spellbind any man alive, whoever comes their way. Whoever draws too close, off guard, and catches the Sirens' voices in the air no sailing home for him, no wife rising to meet him, no happy children beaming up at their father's face. The high, thrilling song of the Sirens will transfix him, lolling there in their meadow, round them heaps of corpses rotting away, rags of skin shriveling on their bones.*[117]

φ φ φ

The lure of the Sirens and the tragic end of the sailors resembles the mating of octopuses, which leads to a tragic end for both the male and the female. Or in the words of King Solomon, who is thought to have written the Song of Songs, "for love *is* strong as death; jealousy *is* cruel as the grave: the coals thereof *are* coals of fire" (Song of Songs 8:6).

An octopus lives most of its life in solitude. Mating is a rare event, one that also appears to be fatal. Before mating, octopuses conduct a courtship in order to consider mate quality and species (interspecies mating occurs, but the result is infertile, at best). In males, one of its eight arms is specialized to transfer sperm from its reproductive organ into the cavity on the female's mantle. During mating, the male deposits the sperm carefully, because a hungry female may prey upon her mate.

Female octopuses can store sperm from multiple mating partners, and occasionally use sperm from different males to fertilize a single clutch of eggs, a process known as multiple paternity. Some scientists believe that sperm competition mechanisms determine which sperm will eventually fertilize the eggs. Others suggest the use of a cryptic female choice, in which the female determines which male octopus will father her offspring by using physical, anatomical or chemical mechanisms to control a male's success in fertilizing. In some cases, the female octopus lays a single clutch with multiple paternity, thereby increasing the genetic variation and perhaps also the fitness and probability of successful reproduction.

117 Homer (1997), *The Odyssey*. Translated by Fagles, R. Penguin Books.

For many adult octopuses, mating means the beginning of the end of their lives, as both the female and male die afterwards. In some species, the male dies a few weeks after mating. He ages rapidly, which means gradual dementia-like deterioration of functional characteristics that lead to death. The female goes through a similar experience, but only after her eggs hatch, during a period that may take up to ten months. She will never meet her offspring, and dies along with all the knowledge she acquired during her life. This phenomenon is called semelparity, meaning a single reproductive episode before death.

There is scientific controversy regarding the evolutionary reasons behind this rapid, programmed death mechanism following reproduction. Experiments have shown that death is controlled by an endocrine organ in the octopus's optic gland. If the gland is removed, the female resumes eating and mating after the eggs have been laid, just as she did before laying the eggs. This means that death is probably not the result of a hunger strike induced by the female's brooding, but the result of an active, adaptive death program attributed to the endocrine system.

The benefits of controlled or programmed death, which is an embedded mechanism in cells, are well known; for example, the controlled destruction of infected/damaged cells, or the programmed death of cells during embryonic development. However, it is not clear why it appears at the animal level (it also appears in Pacific salmon and small marsupials, for example), and especially after such a short lifespan. In cases where the female preys upon the male, it can be explained by increasing female survivability, and as a result, the reproduction process, which is also the goal of the dead male. In other cases, death might be explained by the overproduction of sex steroids, which results in the consumption of most body resources at the expense of the immune system.

One of the evolutionary reasons for the inevitability of death is increasing high fitness in the early life stage, at the expense of survival at a later stage. This makes sense because most creatures end their life in "unnatural" circumstances, such as being eaten by other predators.

Researchers have suggested that the loss of the octopus's external shell caused a dramatic increase in predatory pressure. This increase resulted in shortened life expectancies and fast life histories, characterized by the emergence of advanced intelligence and semelparity.[118] Short life histories and semelparity go together in species with high extrinsic mortality because the high investment of all body

118 Intelligence in large-brained vertebrates might have evolved through independent socioecological pressures and slow life histories. Cephalopods, by contrast, have short life histories and live in simple social environments.

resources in a single fruitful mating event, which has the cost of mortality, seems like a viable strategy when chances of survival in adulthood are low.

Controlled death imparts some fitness to octopuses; it is a means of increasing reproduction success because of the benefits of producing more offspring, and producing them quickly. It acts in the opposite direction to most adaptations that are aimed at escaping death. Whatever the reason for the controlled death of octopuses, it exemplifies, in an extreme manner, the integral role of death in the circle of life.

‹ Mating day octopuses. Eilat, Israel.

‹ Mating blue-ringed octopuses. Lembeh Strait, Indonesia.

Nudibranch Mating

All nudibranchs are hermaphrodite, meaning that adults have both female and male reproductive organs. The male sex organ is stored in a sac on the right side of the head, next to the rhinophore, while the female sex organs are just above the mid-point of the body on the right side. Individuals can fertilize each other and separately lay the fertilized eggs afterwards. The mating position is right side to right side (head to tail), so the sex organs are aligned. This is one structural reason why nudibranchs are not capable of self-fertilization.

>> The reproductive organs of this Elisabeth's chromodoris nudibranch are located on the upper right side. Nudibranchs are hermaphroditic, meaning each individual has organs of both sexes. Puerto Galera, Philippines.

> An Anna's chromodoris lays eggs next to a small ringed favorinus on a gold-mouth sea squirt. The bright colors of many poisonous nudibranchs serve as a warning against potential predators. Verde Island, Philippines.

> Nudibranch eggs. Lembeh Strait, Indonesia.

>> Spanish dancer eggs. The Spanish dancer is a large nudibranch that can grow to a length of up to 60 cm. Sinai Peninsula, Egypt.

Seahorse "Feminism"

Like many vertebrates, the female seahorse produces the eggs and the male produces the sperm. However, unlike many vertebrates, it is the male who becomes pregnant, carrying the embryos until they are born. At the first stage, the female deposits the eggs in the male in a womb-like brood pouch. The male then fertilizes the eggs, keeping them for a brooding period of up to 30 days. During this period, the male is responsible for protecting the eggs and providing them with optimal conditions.

While most seahorses hold the brood with their young in a kangaroo-like pouch on their tail, pygmy seahorses nurture the developing youngsters in a pouch located close to their trunk. At the end of the brooding period, the male releases hundreds of juveniles (some observations indicate even more) and immediately after that, is ready to mate again. Males of most seahorse species mate more than once during the breeding season, and most of them appear to be monogamous, at least within a single breeding cycle and sometimes throughout the breeding season.

˅ Velvet ghost pipefishes are closely related to seahorses, although unlike seahorses, it is the female who carries the eggs. Lembeh Strait, Indonesia.

< A thin ghost pipefish mimics a floating seagrass. Lembeh Strait, Indonesia.

< Ornate ghost pipefish (*Solenostomus paradoxus*). Lembeh Strait, Indonesia.

> Cuttlefish eggs.
Lembeh Strait, Indonesia.

> Anemonefish eggs
before they develop.
Lembeh Strait, Indonesia.

Mandarinfish Mating

Mandarinfish are stunning, benthic dwellers that inhabit the Pacific Ocean, hiding between coral branches or in piles of broken dead corals. Watching their mating ritual is an exceptional experience. Shortly before sunset, at dusk, the females and males start to gather, moving restlessly until they find a mate. Once a pair is determined, they conduct a short courtship dance, aligning themselves belly-to-belly and slowly starting to rise. In a particular moment at the peak of their ascent, about one meter above the reef, they simultaneously release a cloud of sperm and eggs. Immediately after that, they separate and disappear beneath the branches of corals. This process can last half an hour, with several breeding cycles, mostly conducted by the strongest males. Puerto Galera, Philippines.

>> Mandarinfish mating.
Puerto Galera, Philippines.

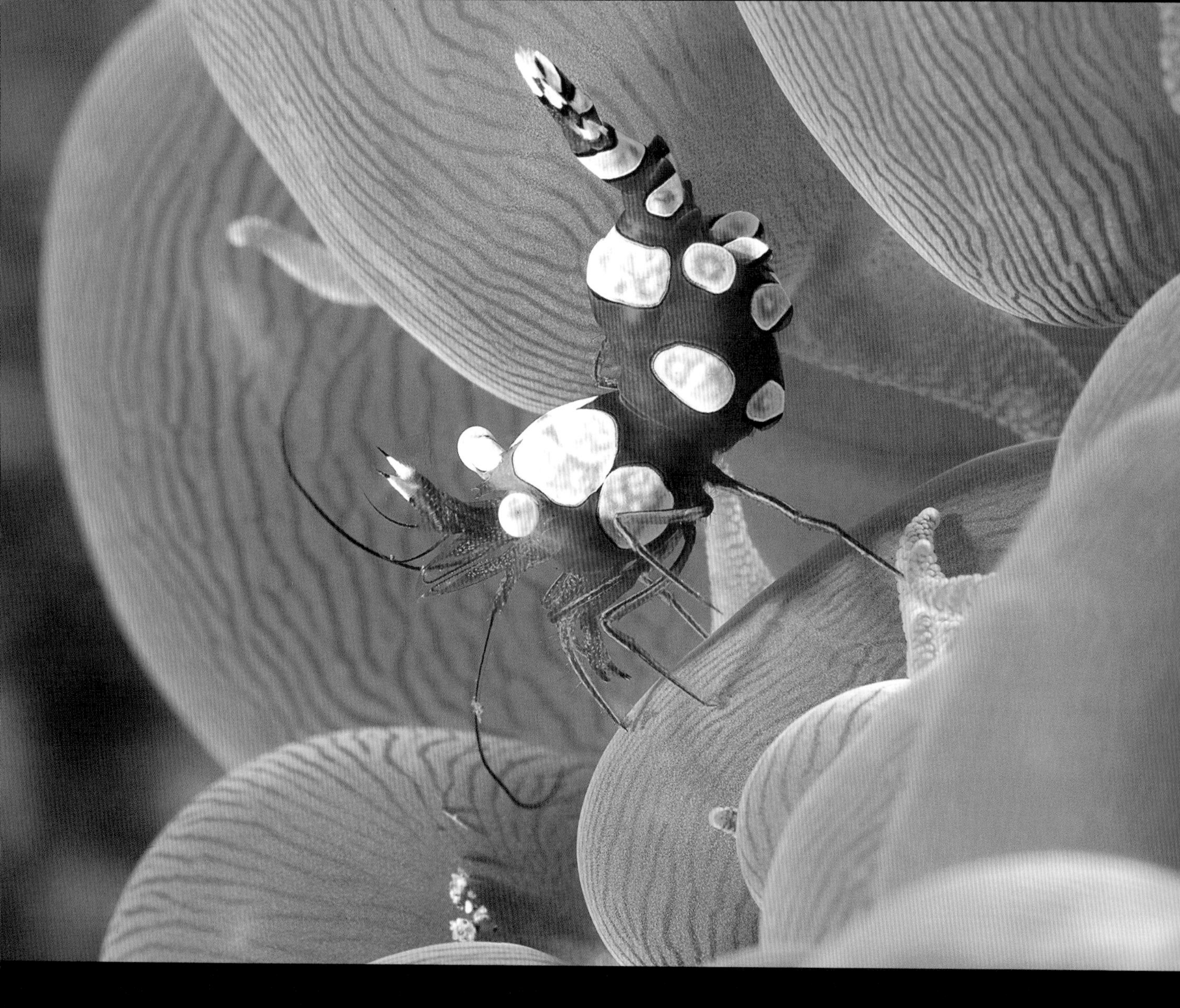

"Through no fault of our own, and by dint of no cosmic plan or conscious purpose, we have become, by the power of a glorious evolutionary accident called intelligence, the stewards of life's continuity on earth. We did not ask for this role, but we cannot abjure it. We may not be suited for such responsibility, but here we are."

Stephen Jay Gould, *The Flamingo's Smile: Reflections in Natural History*, 1985

CHAPTER 8

Evolutionary Themes

« Squat shrimp usually live in commensal symbiosis with various species of sea anemones. Bunaken, Indonesia.

The revolutionary theory of evolution, co-discovered by Charles Darwin and Alfred Russel Wallace, turned science upside-down. It has impacted Western thought to such an extent that American philosopher Daniel C. Dennett (1942–) described it in the following manner: "If I were to give an award for the single best idea anyone has ever had, I'd give it to Darwin, ahead of Newton and Einstein and everyone else." Dennett is referring to Darwin's success in explaining species' evolution without the concept of divine intervention.

A typical visit to a coral reef raises an infinite number of questions about animal behavior, morphology, coloration and more. Evolution theory provides us with the broadest explanatory power concerning the reason behind the diversity of animal behaviors and traits. However, many of the "intuitive" answers to evolutionary questions are the result of misconceptions about the explanatory framework of evolution theory, as well as an overweighted functionalist approach which asserts that all adaptations result from cumulative selections to meet functional ends.

Ever since the theory of evolution was published 160 years ago, scientists have argued about whether and how the theory works. Since the end of the nineteenth century, the concept of evolution has moved firmly away from what was published by Darwin and Wallace, just as physics had done two centuries earlier. In other words, the contemporary perception of evolution theory is far from Darwin's original work, much like contemporary physics is far from Newton's original ideas.

Today's evolution theory is a broader concept called extended evolutionary synthesis (EES), emphasizing a wide variety of inheritance factors. Some would even refer to this new, extended theory of evolution as a paradigm shift. EES is not characterized by giving excess importance to the genetic inheritance factor. It is a broader theory that includes complementary inheritance mechanisms such as epigenetics, behavioral and symbolic inheritances (known as inclusive inheritance), as well as evolutionary development (Evo-Devo), plasticity and niche construction.

The new framework emphasizes the importance and influence of the environment on an organism's phenotype.[119] It deals with questions of the demarcation of the unit of natural selection, both at the individual and the group levels. In other words, what are the borders of an organism (see Chapter 2)? In its broader meaning, evolution theory asserts that there is more to heredity than genes, and that some hereditary variations are non-random in origin. Moreover, it states that some acquired information is inherited, and can result from instruction as well as natural selection. In the following paragraphs, I exemplify some of the concepts that diverge from the "just-so story" of evolution.

The success of evolution theory brought many people to attribute an evolutionary explanation to all types of organism traits. This evolutionary worldview was named by evolutionary biologists Stephen Gould and Richard Lewontin (1929–2021) as biological adaptationism (functionalism) or the Panglossian paradigm. In their influential article entitled "The Spandrels of San Marco and the Panglossian Paradigm: A Critique of the Adaptationist Programme," they significantly changed the discourse of evolutionary biology, warning against confusing function with adaptation. In other words, they show that the existence of a trait does not prove purpose.

The term Panglossian paradigm was named after Dr. Pangloss, the pedantic old teacher in Voltaire's satirical novel *Candide*. Pangloss believed that "all is for the best in this best of all possible worlds," meaning there cannot possibly be an effect without a cause. He adhered to this optimistic worldview despite experiencing great suffering and cruelty. The irony is that this worldview, adopted by evolution theory's adherents in the form of adaptationism, is consistent with the view of German philosopher and mathematician Gottfried Leibniz (1646–1716), who asserted that the world in its present state is the best world that God could have created. Leibniz tried to justify God's acts on Earth, in light of the existence of evil and the world's imperfections.[120] Voltaire (1694–1778), a notable thought leader of the Enlightenment, wrote *Candide* as a satirical criticism against the Catholic Church and an attempt to ridicule Leibniz's worldview.

With regard to adaptationism, Gould and Lewontin referred to the work of German paleontologist Adolf Seilacher (1925–2014), who put forward the idea that

119 A phenotype is a set of observable physical characteristics of an organism, such as morphology, developmental process and behavior. The phenotype is mainly derived from the genotype expression mechanism and the influence of the environment.

120 This is known as theodicy, the vindication of God and defense of God's goodness and omnipotence in the face of evil.

most manifestations of mollusk color patterns are probably non-adaptive and the result of architectural constraints: for example, the color patterns that are not visible because the clams who possess them either live buried in sediments or remain covered by their shell's outermost layer, so the colors cannot actually be seen.

That is to say, not all of an organism's traits can be explained as adaptations; some of them are by-products of historical adaptations, evolutionary mistakes, the result of other constraints, or a shift in the function of a trait during evolution. The conclusion from this approach is that not every phenomenon observed in the reef can be easily explained in a simple, functional manner.

French biologist François Jacob (1920–2013) stated, in his essay "Evolution and Tinkering," that evolutionary innovation (the emergence of novel form and function over time) is different from the optimal engineering process and occurs primarily through a process of tinkering. Tinkering is an opportunistic process in which a "machine" is composed of random combinations of available elements. An engineer works according to a plan based on a defined goal and uses material that has been designed explicitly toward that end. Unlike engineers, evolution works without a clear plan, and with a higher level of constraints regarding the materials and available elements. The evolution of flatfishes' asymmetry, for instance, discussed in the next chapter, is an excellent example of tinkering.

Jointly with paleontologist Elisabeth Vrba (1942–), Gould is also renowned for introducing the evolutionary term exaptation. Unlike adaptation, which is any feature that promotes fitness and is shaped by natural selection for its current role, exaptation is "traits that were adapted for one evolutionary function, but were later co-opted (but not selected) to serve a different role."

An example of exaptation can be found in many fish species that produce sounds for social communication. Those sounds evolved from existing anatomical structures into complex effectors for courtship, without significant changes to their body plan.

Another two examples of exaptations can be seen in sea turtles. The first is their use of flippers for prey capture and processing; the second is the use of their shells for protection. Sea turtles use their flippers for locomotion in water (as swimming paddles) and to move on land (for example, when females go onto the beach to lay their eggs, or in the race of hatchlings back to the sea). It has also been observed that sea turtles use their flippers to manipulate and handle food. According to one perspective, unlike many terrestrial animals, marine tetrapods such as sea lions, whales, penguins and sea turtles were not adapted to use limbs to handle food, but for locomotion. Thus, such a behavior can be considered an exaptation of limbs since the flippers originally evolved for locomotion. It was only at a later stage that they were used to process and handle food.

The second example of exaptation is the functionality of turtle shells. It was recently suggested that the shell of ancient turtles did not evolve for protection but as an adaptive response to fossoriality (its origin was a robust forelimb digging mechanism). The evolutionary process of broadening the rib (around 50 million years before the completed shell) was an adaptation that provided an intrinsically stable base upon which to operate such a robust forelimb digging mechanism. Consequently, the broadening of the rib adaptation served as a platform for the turtle shell's origin. It has also been suggested that fossoriality probably helped turtle ancestors survive mass extinctions.

To summarize, adaptationism is a compelling scientific method of inquiry that has led to impressive achievements. The order in nature is a consequence of natural selection, which governs the most important aspects of trait evolution. Nevertheless, the value of criticism from the adaptationist perspective emphasizes the fact that not all of an organism's traits can be explained by their functions; some of them might be a by-product of evolutionary mistakes that had little effect on fitness, a relic of an ancient trait that is in the (long) process of disappearing, the result of other constraints, or a shift in the function of a trait during evolution.

>> ⌄ Beautiful giant clam mantle patterns. Puerto Galera, Philippines.

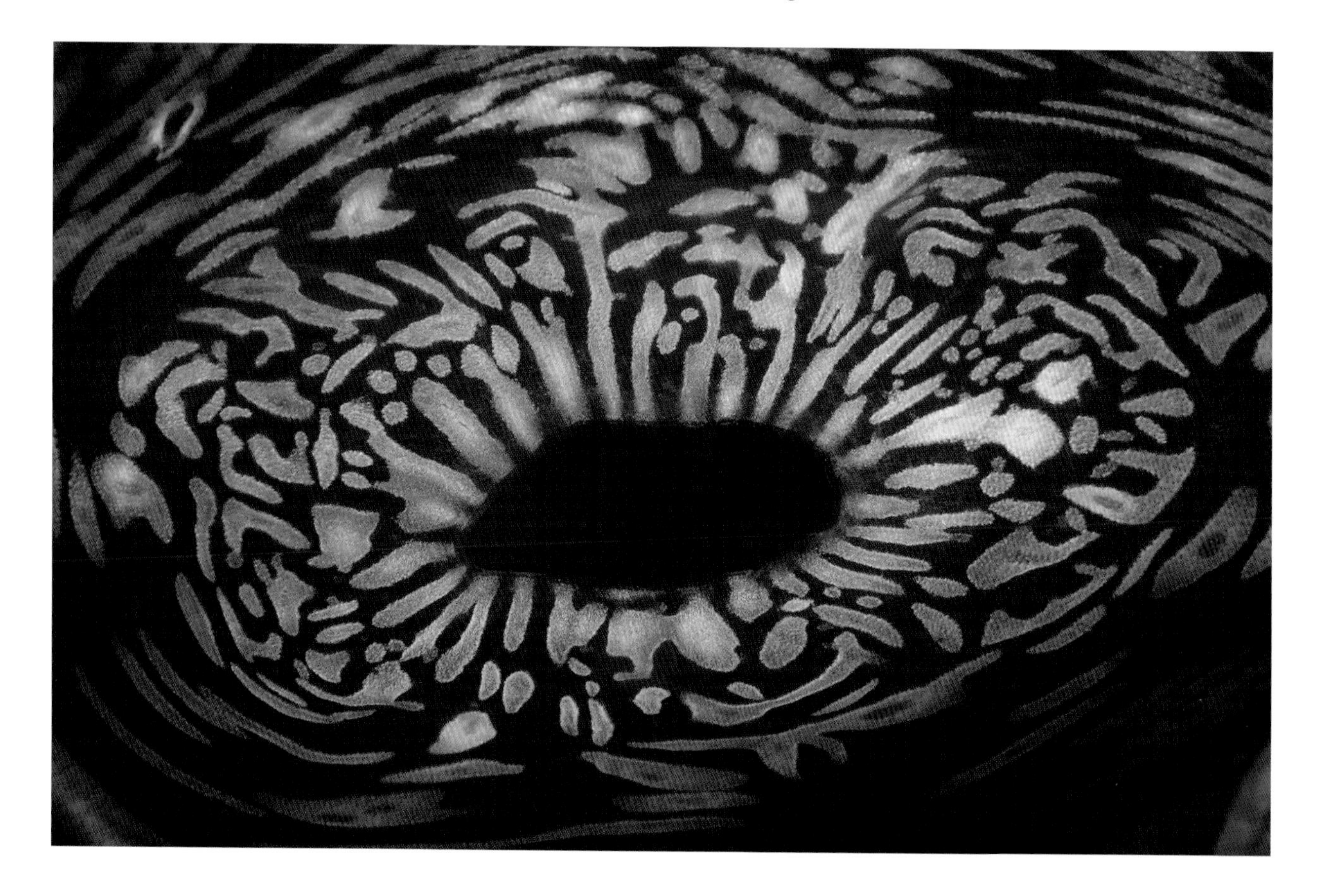

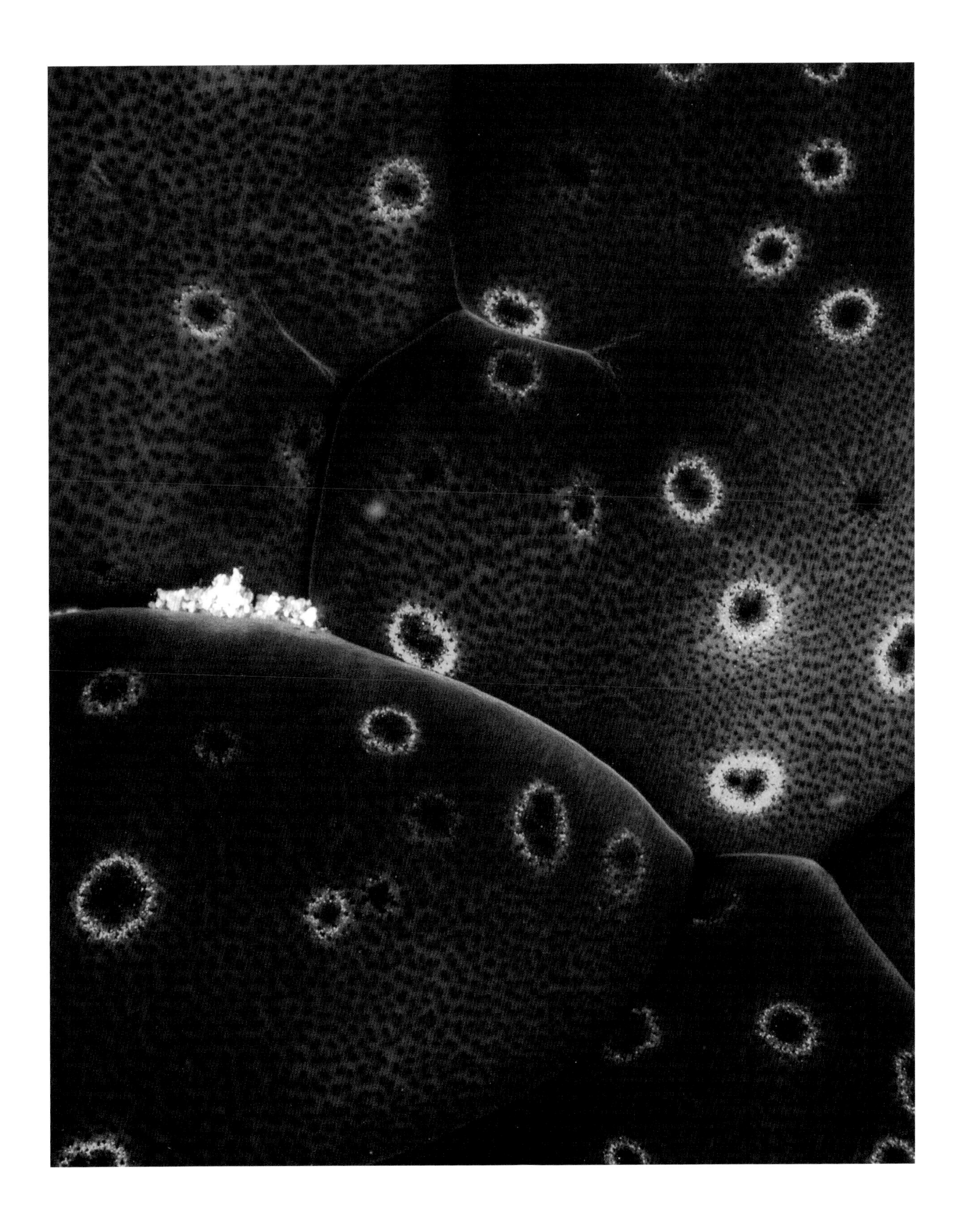

Cephalopods and Animal Consciousness

The waters surrounding Eilat, Israel, are perfect for diving. Maybe not as rich as the Coral Triangle, but very comfortable and beautiful. The blazing sun and the combination of reddish mountains and deep blue water induce relaxation and calmness. During my dives here, I have encountered many octopuses, usually of the common day octopus species. One of these encounters is engraved in my memory, as it was the first time that the question of marine animal consciousness became concrete for me.

In December 2011, I went south to dive at Coral Beach, one of my favorite diving sites in Eilat. I started to submerge my body, descending while letting the air out of my BCD (buoyancy control device). As usual, I was enveloped by crystal clear water. As I sank, I could feel a change in pressure in my ears. My fins gently pushed the water close to the sandy bottom. I was in complete natural buoyancy, and only a slight movement or inhale/exhale of air from my lungs was required to move upwards or downwards. After a short while, I noticed a mess near one of the coral pinnacles. I spun myself upside-down to get a better view and then relaxed into a horizontal position, trying to see what the mess was all about. I noticed a day octopus and a group of yellow goatfishes, maneuvering around one of the rocks (yellow goatfishes are the common companion of day octopuses; they usually hunt together). I was mesmerized by the octopus's smooth movement, reminded of the grace described by French author Romain Gary (1914–1980), in his book *The Kites*:

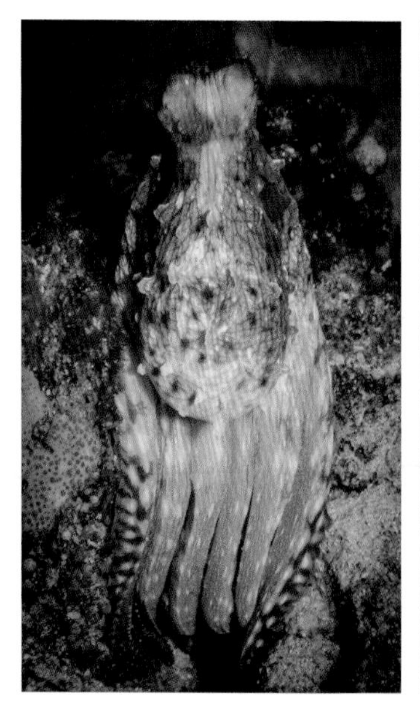

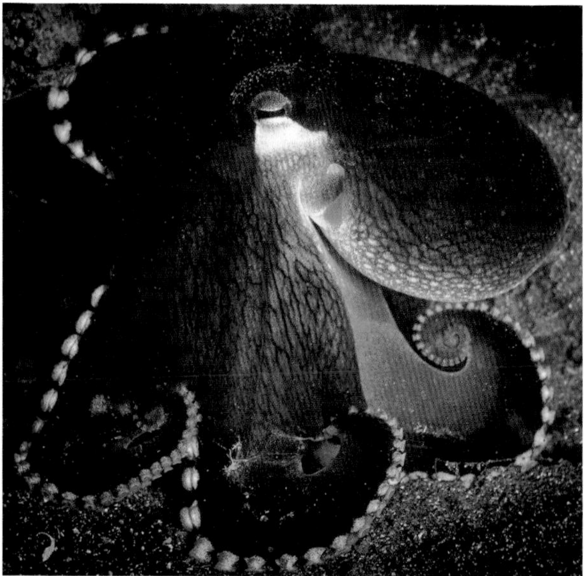

>> Coconut octopus. Lembeh Strait, Indonesia.

> Day octopus gazing at me. Eilat, Israel.

"I had chosen aesthetics as the subject of my exams, and the examiner, who I imagine was worn out after a long day's work, said: 'I shall ask you only one question: What characterizes grace?' I thought of the Polish girl – her neck, her arms, her floating hair – and did not hesitate: "Motion." He gave me nineteen points out of twenty. I owe my *baccalauréat* to love."

Suddenly the octopus stopped moving, moved its body to an upright position, and gazed at me from a very close distance. We looked at each other for a few minutes before I had to ascend and surface. During this mutual gazing, I could feel that he looked at me intelligently, that he was curious about me and about my interest in him. For me, without a doubt, this was a conscious animal and a unique experience, not unlike the one described by ethologist Konrad Lorenz's reflective comment in the book *King Solomon's Ring: New Light on Animal Ways*: "all that I have learned from the books of the library and into the other everything that I have gleaned from the 'books in the running brooks' how surely would the latter turn the scales."[121]

φ φ φ

This section highlights some aspects of the problem of consciousness. My aim is to undermine conventional thoughts about consciousness, and humans' perceived superiority over nature, by softening and broadening the common approach to this complicated and unresolvable issue. I focus on four major points: first, anthropocentrism: the human-centric worldview; second, the hypothesis that consciousness is a pervasive feature of the natural world; third, the suggestion that the primordial marker of consciousness is deeply rooted in the evolutionary history of animals; and fourth, a description of the intelligence of octopuses, which may differ from human intelligence, but still resembles what humans can understand and appreciate. Furthermore, I believe that although this is an elusive discussion, it is an important one, as it has consequences on some of humanity's most burning issues.

121 In this quote, Lorentz uses a passage from Shakespeare's *As You Like It*: "And this our life, exempt from public haunt, Finds tongues in trees, books in the running brooks, Sermons in stones, and good in everything. I would not change it."

Anthropocentrism

One of Western humanity's cognitive biases, some may even say the most central one, is anthropocentrism, the belief that human beings are the most important entity in the universe. This term, which runs like a scarlet thread through the history of ideas, is in line with the concept of human supremacy over nature and is rooted in many people's worldview.

Aristotle developed the idea of *scala naturae* (natural ladder), a hierarchy of the living and non-living world; many animals are characterized by souls yet with different degrees of consciousness. Aristotle believed that both animals and humans possess sensible souls; however, humans are unique because we have the capacity for a rational soul as well. In Aristotle's continuum, humans are at the top, while mollusks (which include octopuses) are close to the bottom. This idea, which was also known as the Great Chain of Being, was also central in Christian thought during the Middle Ages, with God at the top, then angels, then humans, then the rest of the animal kingdom.

In more modern times, philosopher René Descartes (1596–1650) made more significant contributions to the spread of the mechanistic worldview of animals as dumb automatons. This approach was further strengthened by two of the most influential thinkers of the Enlightenment: Scottish philosopher David Hume (1711–1776) and German philosopher Immanuel Kant (1724–1804), both of whom associated consciousness exclusively with humans.[122]

The most fundamental area of disagreement between Charles Darwin and Alfred Russel Wallace was the evolution of the human mind. It is not surprising that evolution theory was developed simultaneously and independently by these two giants, since they were both influenced by their experiences in tropical areas.[123] However, their different personal life stories and backgrounds affected how they saw the fundamental question of humanity's evolution. Darwin argued that human evolution could be explained by natural selection, emphasizing its centrality. In his opinion, there is a continuum between animals and humans regarding the development of the mind. Darwin's idea leads to a firm conclusion that if we have something, other animals have it too, or as Darwin wrote in *The Descent of Man*: "The difference in mind between man and the higher animals, great as it is, certainly

122 Hume identifies the foundation of morality in our emotional responses to the behavior of other human beings. By contrast, Kant identifies it in human rational thinking.

123 Evolution theory was a breakthrough idea, since is contrasted with what most scientists believed at the time: that the world had been created by God, who established divine laws that created perfect adaptations of all organisms to their environment.

is one of degree and not of kind." This means that Darwin did not negate animal consciousness; however, like Aristotle, he saw it as inferior to humans.

On the other hand, Wallace's perspective regarding the human brain was not in line with the adaptationist approach (in most other evolutionary topics, Wallace adhered to adaptationism). Unlike Darwin, Wallace was an autodidact attracted to esoteric, spiritual and revolutionary ideas such as feminism, anti-racism, hypnosis and phrenology.[124] His direct contact with the native peoples of the Amazonas and the Malay archipelago convinced him that they are no different from Western people, a non-conventional idea at that time. He believed that unique human characteristics could not be explained by natural selection and that an external factor, God or "higher intelligence," had to be invoked in order to bridge the gap between the human mind and the animal brain. By the way, this is one of the reasons why creationist religious movements so admire Wallace's doctrine.

One of the prevalent thoughts today, which undermines the traditional anthropocentric view, is that evolution is not driven in a ladder-like manner (i.e. with a hierarchical structure) towards long-term progress, but instead moves towards diversification. Therefore, natural selection is a local adaptation principle and not a general advance or progress with a teleologic nature. Evolutionarily, humans came about through a series of fortuitous and contingent outcomes out of an infinite number of possible trajectories, any one of which could have sent history on a completely different course, one that would not have led to human consciousness. Based on this school of thought, human evolution is just one trajectory, where other "successful" paths could have included moving towards simplicity, as in the case of bacteria and parasites.[125]

Another possible perspective of consciousness, one which differs completely from the anthropocentric point of view, would be from the animal's perspective. Jakob von Uexküll (1864–1944), a Baltic German biologist, outlined an approach that highlighted the animal point of view, calling it *umwelt*. In his view, each organism senses the environment differently, in a manner that cannot be entirely comprehended by human beings. The *umwelt* theory asserts that there are as many world perceptions as there are species.

The same unbridgeable wall between animals and humans was pointed to by Austrian philosopher Ludwig Wittgenstein (1889–1951), when he famously declared in

124 Phrenology is a pseudoscience that asserts that human mental traits can be predicted by measuring the skull structure and bumps.

125 On an evolutionary scale, bacteria are the most dominant form of life due to their ubiquity, quantity, ability to live in widely diverse habitats, and probably biomass.

his book *Philosophical Investigations*, "If a lion could talk, we could not understand him." The essence of what Wittgenstein aimed to say was that since our own experiences are so different from those of a lion, we would fail to understand him, even if the lion spoke our language. In line with Wittgenstein's thought, influential American philosopher Thomas Nagel (1937–) published a notable article entitled "What Is It Like to Be a Bat?" in which he asserted that consciousness is based on subjective experience, and the existing analytical tools cannot grasp the essence of this concept. Different animals have very different ways of obtaining information from the outside world, including forms of long- and short-term memory, learning, sensing, planning and problem-solving.

An example of animals' different worldviews is demonstrated through the mirror self-recognition (MSR) test.[126] In this test, the animal is expected to recognize itself (notice a mark on its body) in the mirror by visually changing its behavior.[127] Animals that successfully pass this test include chimpanzees, elephants, corvids, dolphins and (surprisingly) cleaner wrasses.

It is challenging to imagine fishes and other marine creatures the way we perceive dogs, cats, farm animals and other mammals. However, fishes are sentient creatures, which means they have the ability to feel sensory experiences and, to a limited extent, have a first-person perspective.[128] The fundamental first-person perspective is derived from the ability to experience sensations such as smell, hearing or vision. Behavioral evidence supports the existence of the basic first-person perspective in fishes. Not all fish behavior is the result of mere reflexes or unconscious responses.[129] An example of this is the successful passing of the MSR test by cleaner wrasses, and fishes' ability to identify the time and place of feeding stations.

When I look at my beloved dog Whiskey, I do not doubt that he is a conscious animal with a self-identity that might be based on smell rather than self-portrait. This thought might sound baseless; however, an experiment carried out by Dr. Alexandra Horowitz suggested that dogs, which usually fail the MSR test, can self-recognize their own smell.

---

126 https://youtu.be/Okmkn30DoNU (retrieved March 5, 2021).

127 In human babies, around 65 percent pass the MSR test by the age of 18 months.

128 A low order of the "first-person perspective" is the ability to experience sensations; a high order, such as humans have, is self-consciousness.

129 https://aeon.co/essays/fish-are-nothing-like-us-except-that-they-are-sentient-beings (retrieved March 5, 2021).

< My beloved dog Whiskey.

Basic philosophical approaches to consciousness

Consciousness is an elusive term and for many years, evolutionary biologists avoided coping with it. The difficulty in addressing it is echoed in a phrase attributed to Einstein, "No problem can be solved from the same level of consciousness that created it." In other words, we can never fundamentally and objectively understand the phenomena of consciousness since humans do not have the capacity to introspect their consciousness.[130]

In general, there are two major philosophical perspectives to addressing the question of the origin of consciousness. The first is physicalism, which asserts that consciousness can be explained bottom-up; that is, by physical and biological laws. The second approach is mind–body dualism, which holds that mental phenomena differ from physical phenomena. In other words, mind and matter are distinct, different kinds of entities. These two different schools of thought are at the heart of a longstanding philosophical debate known as the "mind–body problem." So far, those two perspectives have not yielded a real understanding of the causes and source of consciousness and it seems likely that the debate about them is not close to resolution.

130 This corresponds with Gödel's incompleteness theorem, named after German-Austrian mathematician and philosopher Kurt Gödel (1906–1978). Gödel's theorem expresses that it is impossible to prove that a system of axioms is consistent without using a different set of axioms. In other words, human consciousness can only be understood if it is explored by an entity with consciousness that is separate from, and superior to, that of humans.

Nevertheless, there is a third approach that ontologically rips the ground right out from under physicalism and the mind–body dualism. This is panpsychism, meaning everything has a mind, which has ancient roots in animist religions, pre-Socratic thinkers, and Plato, and is an integral part of the Tibetan Buddhist tradition that emphasizes the continuity of consciousness across life.

Panpsychism has many variations. Basically, it postulates that mentality is an intrinsic property and pervasive feature of the natural world, just like the force of gravity. It asserts that micro-level entities have mentality, which is found in all things throughout the material universe.[131] Based on panpsychism, subjective experience and, eventually, human consciousness are made from integration between conscious particles that form complex structures.

Panpsychism may sound ridiculous at first; however, there is a long list of highly distinguished scientists and philosophers, including contemporary scholars, who support it. These include Dutch philosopher Baruch Spinoza (1632–1677), American philosopher and psychologist William James (1842–1910), British philosopher Bertrand Russell (1872–1970), Jesuit paleontologist Teilhard de Chardin (1881–1955), physicists Freeman Dyson (1923–2020) and David Bohm (1917–1992), and neuroscientists Christof Koch (1956–) and Giulio Tononi (1960–), to name a few.

There are early signs of gaining academic credibility and the support of leading scholars in panpsychism. An example is Professor Philip Goff, who recently asserted that in the case of panpsychism, "scientific support for a theory comes not merely from the fact that it explains the evidence, but from the fact that it is the best explanation of the evidence." Others support the theory based on various humanistic and scientific disciplines. One of them is the nature of matter as described by quantum physics. In the words of the great physicist David Bohm, "That which we experience as mind ... will in a natural way ultimately reach the level of the wavefunction and of the 'dance' of the particles. There is no unbridgeable gap or barrier between any of these levels ... It is implied that, in some sense, a rudimentary consciousness is present even at the level of particle physics."

Additional arguments that support this approach are the fundamental requirement for continuity between matter-life and consciousness, and of course, the unique elusive nature of consciousness that cannot be described by the other perspectives.

131 https://plato.stanford.edu/entries/panpsychism/#DefiPanp (retrieved March 5, 2021); https://daily.jstor.org/panpsychism/ (retrieved March 5, 2021).

The recent revival of interest in panpsychism has several possible consequences on human thought. The first is its ability to explain how life and consciousness evolved from inanimate and non-conscious matter. The second is that panpsychism paves the way for the possibility of attributing different forms of consciousness to non-human entities and undermines the anthropocentric view that, in its extreme, asserts that only rational agents gain membership in the moral community.

The evolution of the sensitive soul

An interesting materialistic hypothesis regarding the evolution of consciousness was recently suggested by Professor Simona Ginsburg and Professor Eva Jablonka in their 2019 book entitled *The Evolution of the Sensitive Soul*. In this book, they attempt to address the fundamental question of the cognitive marker of the presence of consciousness. They succeed in broadening the meaning of the term consciousness and establish the foundation for a significant change in the way humans perceive animal consciousness. In their opinion, the "subjective experience" is the paradigmatic process that is identified with conscious experience and is, therefore, equivalent to the terms sentience and consciousness. They claim that the evolutionary qualifier for minimal consciousness is unlimited (open-ended) associative learning (UAL). UAL enables animals to attribute motivational value to a composite of stimulus or actions and use it as the basis for future learning.

Ginsburg and Jablonka posit that this primordial UAL system emerged during the Cambrian period about 542 million years ago. During this period, bilateral symmetry, the primitive nervous system and the origin of associative learning probably led to an evolutionary arms race. The Cambrian period was also characterized by the intensification of stress, disease and suffering, which were also drivers for the evolution of UAL. According to the authors' hypothesis, a UAL system, which eventually entails consciousness, emerged in a feedback manner with increased fauna diversification. Moreover, it might have also been one of the major driving forces behind the Cambrian explosion.

Two prerequisites for UAL are a brain and a central nervous system. Sponges, a major reef dweller, do not have any nervous system at all; cnidarians, which appeared later in evolution, have a dispersed, primitive one, which is not an entirely "blank slate." Observations of jellyfishes and sea anemones have shown early signs of learning systems (still non-associative) that enable them to adapt more effectively. In a very simplistic non-associative manner, cnidarians can modify their behavior as a result of past experience through sensitization and adjustments.[132] This means that cnidarians might be precursors of primordial early consciousness.

In addition to being challenging and intriguing, the above hypothesis has moral consequences, by extending the border of consciousness to include animals who, until now, have not been perceived as conscious. Ginsburg and Jablonka propose that "consciousness can be attributed not only to all vertebrates but also to some invertebrates in addition to cephalopods." Secondly, it identifies early primordial signs for UAL, hundreds of millions of years ago, and states the continuity of consciousness evolution and the connection between humans and animals with other types of consciousness.

To sum up, it seems that the anthropocentric view regarding consciousness is narrow, not to say baseless. The advent of consciousness is still under scientific and philosophical debate, and the current prevailing paradigm is not even close to grasping and understanding it. Many animals have subjective experiences, and probably a different type of consciousness that started evolving more than 500 million years ago.

Octopus consciousness

Octopuses, which are frequently seen in coral reefs, appear in many discussions of animal consciousness. This is not coincidental, since octopuses are an exceptional evolutionary "experiment" that demonstrates the unique trajectory of a sentient animal, one that is utterly different from that of humans, yet still close enough to what humans can understand and appreciate.

Evolutionarily, the common worm-like ancestor of humans and cephalopods (octopuses, cuttlefishes and squids) lived perhaps 500–600 million years ago, and since then, the two lineage trajectories have never crossed. Our ancient common ancestor probably had none of the consciousness skills that both octopuses and humans possess today.

Cephalopods belong to the phylum *Mollusca*, which also includes animals such as clams, bivalves and sea slugs. Of all marine invertebrates, only cephalopods, and some would say certain species of mantis shrimps, developed a complex nervous system. Cephalopods have invested in a "costly," distributed nervous system, as

132 There is an indication (only one) for associative learning in cnidarians. In research conducted by Haralon and Groff, a small sample of sea anemones learned to contract their tentacles when presented with light that predicted a shock. Other experiments to condition cnidarians have failed.

< Some coconut octopuses use tools such as clam shells for concealment and defense. Lembeh Strait, Indonesia.

the signaling between neurons and different organs requires a continual flow of chemicals to support the electric communication system. Besides having a complex central nervous system (a brain) with learning capabilities and long-term memory, octopuses have eight arms that are controlled by many neurons (two-third of their body's total neuron count). These tentacles can be guided from the central brain and are likely capable of autonomous reflex and light sensitivity. However, unlike the octopus's central brain, the arms are not capable of associative learning.

In his book *Other Minds: The Octopus, the Sea, and the Deep Origins of Consciousness*, Australian philosopher Peter Godfrey-Smith (1965–) suggests the following metaphor for the interaction between the octopus's brain and its arms: "In the octopus's case, there is a conductor, the central brain. But the players it conducts are jazz players, inclined to improvisation, who will accept only so much direction. Or perhaps they are players who receive only rough, general instructions from the conductor, who trusts them to play something that works."

Octopuses have uniquely proved to be intelligent,[133] however unlike humans, and as asocial animals, all their acquired knowledge is learned independently and dies with them. They have done very well in human-designed tests such as learning by observation from other octopuses, problem-solving, playfulness and future planning,

133 Intelligence can be defined as the capacity to apply information and cognition for successful problem solving.

not just in their natural environment but also when faced with artificial tasks. They are exploratory animals that forage for food in diverse methods that require flexible thought and advanced manipulation skills.[134] They do not wait in ambush like many other predators but actively search for food by constantly moving.

One example out of many octopus skills is the ability of giant Pacific octopuses to distinguish between different humans, even when they are wearing the same uniforms. Another example can be found in the behavior of coconut octopuses, that assemble pairs of coconut shells or clams as a portable shelter, carrying them to an appropriate place and building a shelter from them for future use.[135] Furthermore, some scientists have tentatively hypothesized that octopuses might dream while sleeping.[136]

Evolutionarily, the two primary driving forces behind large brains were probably a response to challenges associated with finding and processing food (ecological intelligence) and group living demands, such as maintaining complex and enduring social bonds, deception and cooperation (social intelligence). In the case of vertebrates, the central hypothesis regarding large-brained intelligence is the coevolution of intelligence with long life histories.

Cephalopods, on the other hand, evolved in a different evolutionary trajectory. They developed large and complex nervous systems, advanced sentient capabilities, and high intelligence, that coevolved with short life histories.

It seems that the primary function of octopuses' sophisticated intelligence is foraging for food, in a hostile environment, with a limited defense mechanism (a result of the loss of their shell 275 million years ago). The significant increase in predatory risk was mitigated by their iconic camouflage capabilities and intelligent anti-predatory strategies. Social intelligence probably only played a secondary role in intelligence evolution since complex social bonds are not well-developed in octopuses.

There is a growing consensus about the high level of sentience in cephalopods and other invertebrates.[137] In a comprehensive meta-analysis (drawing on over 300 scientific studies) published in 2021 by LSE consulting, researchers evaluated evidence of sentience in two groups of invertebrates: cephalopods (including octopuses, squid and cuttlefish) and decapod crustaceans (including crabs, lobsters and crayfish). They found strong evidence of sentience in octopods and, to a lesser extent, in decapod

---

134 An octopus takes a fish from a jar: https://youtu.be/_xfDYs_6rUk (retrieved January 17, 2021).

135 https://youtu.be/BFda1MZ54G4 (retrieved January 17, 2021).

136 https://youtu.be/ovKCLJZbytU (retrieved January 17, 2021).

137 Sentience is defined as the capacity to experience feelings of pain, pleasure, hunger, thirst, warmth, joy, comfort and excitement.

crustaceans – results that may well have practical implications. The report's primary recommendation is "that all cephalopod molluscs and decapod crustaceans be regarded as sentient animals for the purposes of UK animal welfare law." Eventually, I believe that many invertebrates will be included within our moral demarcation line, and this may have a significant positive impact on how we see our relationships with nature.

The importance of the discussion about consciousness

The discussion of consciousness is crucial since it raises the question of the demarcation of the moral circle, and whether we should include cephalopods and other animals within it.

Science and philosophy are unable to provide conclusive proof regarding the level of consciousness of non-human creatures. However, there is evidence that octopuses and fishes often exceed other vertebrates when it comes to perception and cognitive abilities. Thus, it would be fair to say that we cannot rule out the possibility of developed consciousness in many animals as part of the "space of possible minds" that is ubiquitous in nature. The foregone conclusion is that we should grant octopuses, fishes, and other animals the same protection level as any vertebrate.

Extending the moral circle and including within it a wide variety of animals may change our perspective regarding the place of humans in the world. Understanding that we are just one component of an immensely interconnected web of life can serve as a call to humanity for action regarding the chronic problems of environmental conservation.

I would like to conclude this section with a quote from Albert Einstein, from a 1972 article in the *New York Times*, that describes the importance of extending the line of moral demarcation. Einstein exchanged letters with a rabbi who had trouble comforting his 19-year-old daughter following the death of her younger sister:

> *A human being is a part of the whole, called by us "Universe," a part limited in time and space. He experiences himself, his thoughts and feelings as something separate from the rest – a kind of optical delusion of his consciousness. This delusion is a kind of prison for us, restricting us to our personal desires and to affection for a few persons nearest to us. Our task must be to free ourselves from this prison by widening our circle of compassion to embrace all living creatures and the whole of nature in its*

beauty. Nobody is able to achieve this completely, but the striving for such achievement is in itself a part of the liberation and a foundation for inner security.

Origin of the Octopus: Terrestrial or Cosmic?

The idea that life occurs outside of Earth is not new, and is intertwined with hypotheses regarding the extraterrestrial source of life on Earth. This idea has ignited the imaginations of scientists and philosophers throughout history, and still plays a part in popular culture and, to a limited extent, science. Some of the most profound questions we ask are what is the source of life on Earth? And are we alone?

An example of the popularity of the idea that extraterrestrial forces influenced life on Earth is Erich von Däniken's (1935–) highly controversial bestselling book *Chariots of the Gods*. In it, Von Däniken argued that ancient cultures were influenced and acquired knowledge via contact with aliens ("ancient astronauts"). In his opinion, these contacts explain many cultural puzzles, including how the Egyptian pyramids were built (using advanced alien technology), the mysterious Nasca lines in Peru (alien airfields), and some descriptions from the Book of Ezekiel that Von Däniken attributed to ancient people's descriptions of spacecraft and alien bodies.[138]

This type of belief is sometimes based on a logical fallacy known as *argumentum ad ignorantiam* (argument from ignorance), which means that if there is no agreeable and satisfactory explanation, then the alternative theory is valid. In our case, if existing terrestrial theories about the construction of the pyramids are not persuasive enough, then we should deduce that they were built with advanced alien technologies. Many theories about the influence of extraterrestrial life on Earth are far-fetched and based on false science. However, we need to be cautious with over-generalization. Variations of these theories have been proposed by credible scientists, are based on scientific evidence and interpretations, and present an authentic challenge to the existing terrestrial paradigm.

138 For example, the following verses in Ezekiel 1: "and a brightness was about it ... Also out of the midst thereof came the likeness of four living creatures ... they had the likeness of a man ... and every one had four wings ... and their feet were straight feet..."

φ φ φ

One of these hypotheses is panspermia (Greek for "all seed"), which was mentioned in ancient history, as well as in the second half of the nineteenth century. This hypothesis proposes that microscopic lifeforms can survive extreme space conditions (for example, be trapped in the celestial matter) and be carried and distributed to Earth by space dust, meteorites, asteroids, comets and more.

In 1974, Fred Hoyle (1915–2001), a British astronomer, and Chandra Wickramasinghe (1939–), a Sri Lankan–British astrobiologist and mathematician, argued that lifeforms (for example, bacteria, fertilized ova, seeds and viruses) entered the Earth's atmosphere 4 billion years ago because the conditions on Earth allowed them to thrive. Later, virus-bearing cometary-bolide bombardment events coincided with major mass extinction–diversification and may have been responsible for the genetic novelty necessary for macroevolution, as well as for many significant epidemics throughout history. In Hoyle and Wickramasinghe's (H–W) opinion, viruses, which are among the most information-rich natural systems, can survive with high probability, ionizing the radiation and heat that are part of their journey due to their small size.

The H–W idea has not faded entirely. In 2018, in a controversial article entitled "Cause of Cambrian Explosion – Terrestrial or Cosmic?," a group of 33 scientists, including molecular immunologist Edward Steele and the aforementioned Wickramasinghe, suggested that recent evidence and studies support the H–W hypothesis. In their opinion, this hypothesis is the primary explanation for the Cambrian explosion 542 million years ago, an event known as the Big Bang, that resulted in the divergence and diversification of most animal phyla. The article also argued that the "extended" H–W hypothesis is a significant scientific paradigm shift in understanding the evolution of life on Earth and the origin of many pandemics.

The authors based their hypothesis on a few arguments. The first is phylogenetic analysis regarding the origination of vertebrate retroviruses prior to the Cambrian period.[139] These play a significant role in the genomic shaping process, and were one of the drivers of the Cambrian explosion. The second comes from fossils of micro-organisms found in meteorites, and life-bearing particles from space found in the atmosphere. The authors also propose that intelligent characteristics and the unique DNA structure and transcription mechanism of some cephalopod species also substantiate H–W predictions.

139 Retroviruses are a family of viruses containing RNA, that is duplicated by invading a cell and inserting a copy of the virus's RNA into the cell's DNA.

Can viruses and genetic material endure the conditions in space?

The first question that needs to be asked is what is the likelihood that genetic material will survive space travel, especially in light of the conditions in space, which are very hostile for organisms? Cosmic radiation, close to absolute zero temperatures, vacuum, ionizing radiation, and a lack of oxygen are only a few of the environmental conditions that an organism has to endure in space. However, some animals can tolerate these types of extreme conditions. Tardigrades, for example, also known as "water bears," are tiny eight-legged animals found in the deep sea, the Arctic, the Himalaya mountains, tropical forests, and almost every habitat on Earth. They can survive temperatures in the range of −272°C to 150°C for a few minutes and −20°C for decades, as well as pressure reaching up to 1,200 atmospheres. The question of how they endure such extreme conditions is still open. It has been proposed that they suspend animation so that their metabolism and other life processes come reversibly to a standstill, enabling them to withstand extreme stress conditions. Scientists from Harvard and Oxford who simulate catastrophic space events found that some water bears would survive them. While this does not prove the panspermia hypothesis, it does put the possibility that some forms of life might survive a space journey and catastrophic event on Earth in a different light.

The origin of cephalopods

Associating cephalopods with panspermia is not surprising, since they have always attracted the attention of the scientific community due to their behavioral repertoire, disguise mechanisms, resourceful predation skills, and a wide variety of morphological innovations.

For example, the octopus genome belongs to a subfamily with 500 million years of evolutionary trajectory that shows an amazing level of intelligence and complexity, with more protein-coding genes than humans. Moreover, the genetic divergence of the octopus from its ancestors (the same ancestors as the modern nautilus) is significant, and might not have come only from horizontal gene transfers, a random mutation of existing genes or duplicative expansions.

In general, it is assumed that genetic, blueprinted information is transferred, with high fidelity, from DNA→RNA→protein. In most animals, a change in the protein structure requires a change in the DNA, sometimes by random mutation.[140]

140 In the simplest sense, DNA is a pair of molecules found in each cell's nucleus that contain codes (the "recipe") to produce proteins that are the building blocks of body tissue. Expressing a gene means producing its corresponding

Similar results can be achieved by a process called RNA editing.[141] It was recently discovered that octopuses, along with some squid and cuttlefish species, routinely and extensively program their RNA sequences to adapt to their environment. RNA editing gives organisms a flexible and fast option to express diverse, functionally distinct proteins and enrich genetic information beyond the genomic DNA blueprint, mainly in the nerve system. It sets the base for a mechanism that creates evolutionary novelty, compared with other organisms. One study that looked at possible mismatches between RNA and DNA showed that in octopuses the amount of RNA molecules that were subject to editing is more than 50 percent; this compares with 3 percent in humans and a similar amount in other animals.

Nonetheless, the RNA editing that enables enormous phenotype flexibility comes with a price tag. It slows down the accumulation of mutations and thereby the rate of conventional evolution (at the DNA level). However, by changing RNA without changing DNA, cephalopods could produce different proteins (from the same gene) that work better in different conditions. Furthermore, such changes would be temporary – the creatures could turn them on or off depending on the circumstances.

The enormous qualitative difference in the RNA editing of cephalopods (excluding nautilus) compared to other animals is used to support the H–W panspermia hypothesis. In other words, perhaps cephalopod DNA was carried to Earth by an icy comet or extraterrestrial virus that infected a cephalopod ancestor. Needless to say, this hypothesis counters the accepted scientific assumption: that octopuses gradually developed from nautiloids.

φ φ φ

In a way, the scientific debate about the origin of life corresponds with an anthropological phenomenon known as "cargo cults." Cargo cults emerged in the last century in the South Pacific region of Melanesia, a diver's paradise that includes Australia, Papua New Guinea, Fiji, the Solomon Islands, Vanuatu, New Caledonia, and additional Pacific islands.

---

protein. The manufacturing instructions are transferred to a messenger RNA (mRNA) molecule in a process called transcription, where the end of the process results in a single-stranded copy of the gene. The next stage is manufacturing the protein molecule based on the instructions written in the RNA.

141 RNA editing is the alteration of the RNA sequence in the RNA after it has been transcribed from DNA and before it is translated into protein. The consequences of RNA editing are equivalent to DNA mutations, but without changing the genetic code.

As a response to Western colonialism, mainly during the Second World War, and the introduction of technological goods, some groups of people developed messianic or millenarian belief systems. They perceive Western technological goods such as airplanes, guns, tools and modern food as divine objects, and developed rituals aimed at causing their deities and ancestors' spirits to deliver valuable European and American goods (cargo). Some cargo cult adherents built replicas of airstrips and airplanes using trees and branches, and imitated sounds associated with airplanes to initiate the shipment of valuable goods.

Cargo cults allow anthropologists to observe the creation of rituals and belief systems in real time, and demonstrate how some indigenous people interpret modern phenomena. Some supporters of the pseudoscientific "ancient astronauts" theory mentioned earlier draw an analogy between today's cargo cults and theories that ancient cultures met with aliens and benefited from their advanced technology.

Are supporters of the panspermia theories comparable to members of cargo cults because they both invoke extraterrestrial explanations for terrestrial phenomena? Not quite. In my opinion, suggestions of a non-terrestrial origin of octopuses should not be rejected automatically, but should be considered alongside more traditional scientific explanations. Perhaps these ideas are not mutually exclusive, and can live together in some fashion.

Throughout history, ideas about extraterrestrial life sources have been shaped by prevailing philosophical and scientific concepts. Modern times are no exception; even though the scientific community generally rejects panspermia, the discussion about it has not disappeared. Finding causal explanations to events is a basic primordial human need that provides us with a stable and safe worldview of the environment. Subconsciously, humans feel awful thinking about the possibility that we are alone in the universe. Moreover, it also satisfies our motive to understand the world coherently and consistently. Unfortunately, the fulfillment of these primordial motives sometimes, as in the case of the "ancient astronauts" theory, comes at the expense of critical thinking and rationality.

The Evolution of Animal Eyes: Convergent or Divergent Evolution?

In the sixth chapter of *On The Origin of Species*, Charles Darwin addresses eyes, describing them as "Organs of Extreme Perfection":

> *To suppose that the eye with all its inimitable contrivances for adjusting the focus to different distances, for admitting different*

> *amounts of light, and for the correction of spherical and chromatic aberration, could have been formed by natural selection, seems, I freely confess, absurd in the highest degree.*

It is not surprising that creationists refer to this quote as proof of Darwin's skepticism about his own theory. However, those creationists tend to avoid the next paragraph, where Darwin explains the gradual manner in which eyes developed, from very simple eyes until they reached perfection.

Since the creation of the earliest life form on Earth more than 3.7 billion years ago, sunlight has been the most prominent selective force that influences the evolution of organisms. Surprisingly, within the animal kingdom, most animals did not evolve sophisticated eyes. Out of 35 animal phyla, about one-third have no specific organ for sensing light, one-third have light-sensitive organs, and the remainder are animals with what we consider to be eyes.

Although life has existed for a few billion years, advanced vision systems only appeared half a billion years ago, during the Cambrian explosion. Most of the basic visual structures that exist today also existed during that period. Since then, they have been diversified in an extraordinary variety of ways for specialized tasks.

In his book *In the Blink of an Eye: How Vision Kick-Started the Big Bang of Evolution* (2016), British zoologist Andrew Parker (1967–) argued that eyes played a key role during the Cambrian explosion. Parker claims that the development of sophisticated eyes facilitated animal locomotion, which in turn resulted in an increase in competition and predation capabilities, and was therefore among the major factors behind the burst of animal diversification during the Cambrian explosion.

Coral reefs are full of amazing eye configurations that cover a broad spectrum of vision systems. There are at least ten different, optically distinct, types of eyes. These include camera-like eyes (fishes and octopuses), compound eyes (shrimps), mirror-like eyes (scallops), and the primitive light-sensitive patches on flatworms (a simple eye called an ocellus). All of these eye types are discussed in Chapter 4.

The wide variety of eyes is so distinct that biologist Ernst Mayr suggested that eyes have evolved 40–60 times independently. This type of evolutionary hypothesis is called convergent evolution, where similar features in species of different periods or epochs evolve separately. The idea behind convergent evolution is that natural selection acting on random variations will repeatedly lead to similar solutions to biological problems. The other possible evolutionary trajectory that can explain the diversity of vision systems is divergent evolution, which means that eyes evolved from common descent.

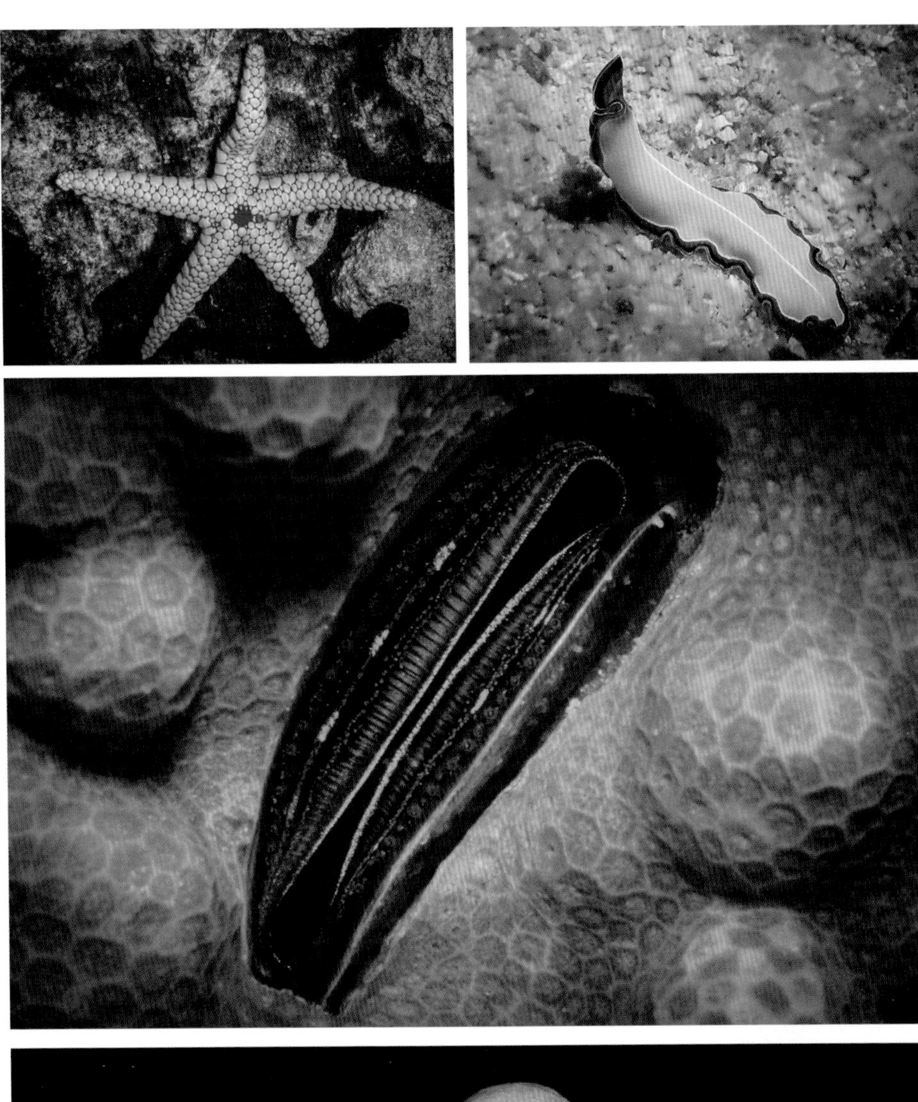

>> Flatworms have a primitive, light-sensitive, simple eye known as an ocellus. Dauin, Philippines.

> The eyes of sea stars (starfish) are located on the tip of each arm. They cannot see colors, but they can detect light and dark. Experiments have shown that shallow-water sea stars use their eyes to find suitable feeding locations in the reef. Eilat, Israel.

> Scallops (for example, iridescent scallops) possess a visual system comprised of up to 200 tiny eyes on the upper and lower edge of the mantle. These unusual eyes do not focus light through a lens; instead, they contain a multilayered concave mirror at the back of the eye that reflects the image onto the retina above it. This structure acts as a focusing lens while doubling the chance of capturing incoming light. The scallop's mirror eye resembles the segmented mirrors of the reflecting telescope invented in the seventeenth century by Isaac Newton and commonly used today (for example, in the Hubble space telescope). Eilat, Israel.

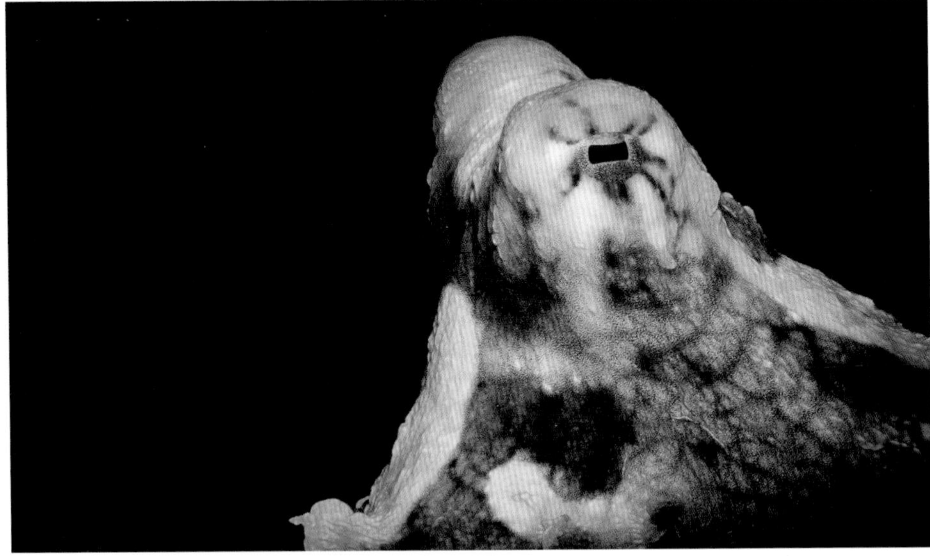

> Day octopus eye with dumbbell-like pupil. Eilat, Israel.

>> Thorny (spiny) seahorse. Lembeh Strait, Indonesia.

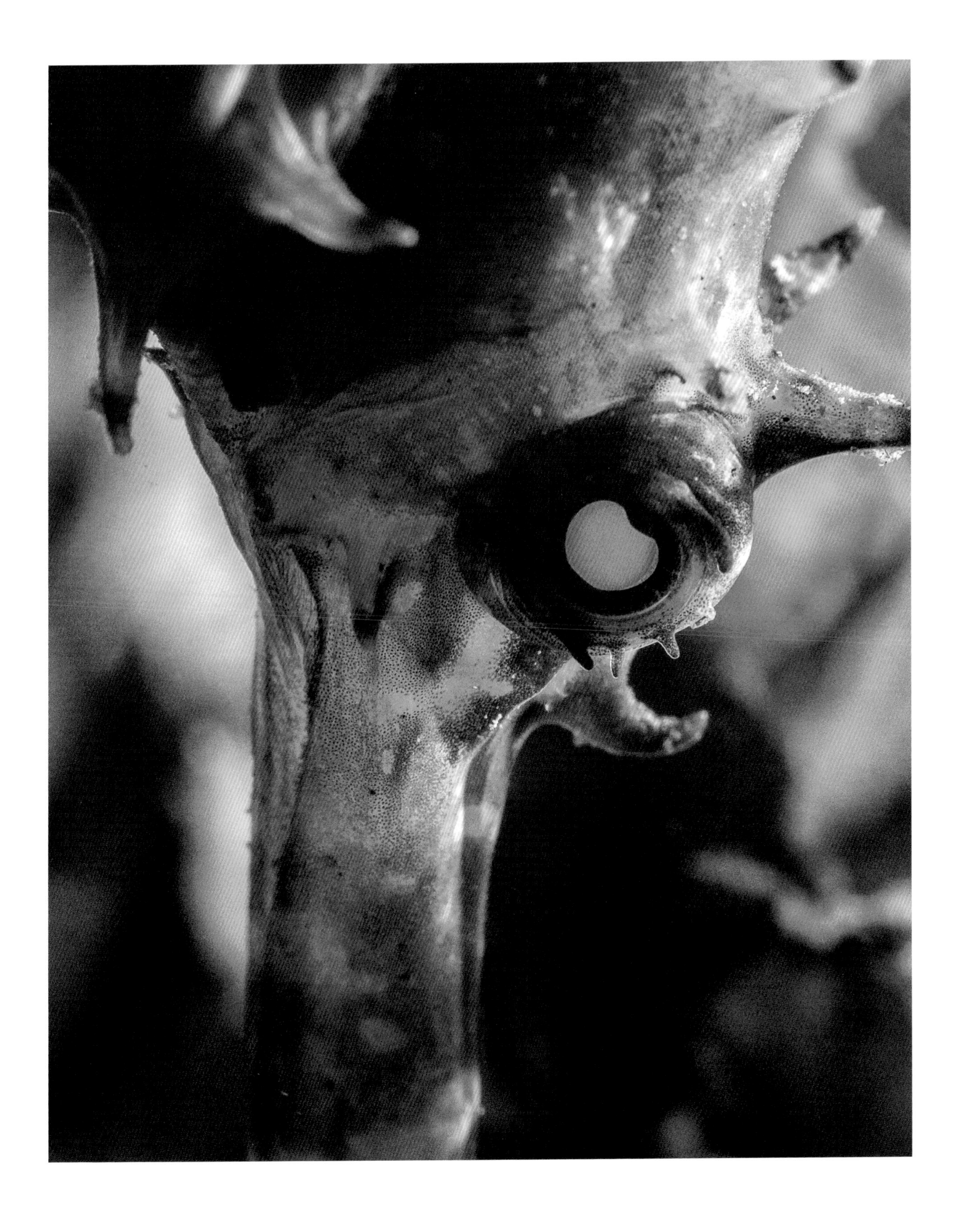

Recent genetic analysis casts doubt on Mayr's hypothesis regarding the independent evolution of eyes, and suggests that eyes evolved only once, followed by divergent, parallel evolution. The eyes of many animal species that evolved in many ways and from different tissues still share deep homologies on the molecular level.[142] These primordial molecular structures, genetic control networks, and physiological functions have repeatedly been used and co-opted for similar purposes in different species.

A good demonstration of this hypothesis is a group of proteins called opsins. These light-sensing proteins are unique to animals and responsible for vision in all animals, from jellyfishes to humans. Based on this hypothesis, all animals share the same biochemical foundations of vision systems, suggesting the existence of common ancestry that might be extremely ancient, possibly even before the split between algae and animals.

The pace of complex eye evolution was probably very fast. A computer simulation built by professor Dan-Eric Nilsson shows the complete development of an eye, from a small flat patch of light-sensitive pigment to a fully functioning camera-like organ, within just 364,000 generations (0.005 percent change per generation), or the same number of years (364,000 years) in a small invertebrate. This simulation may explain the fast development of complex vision during the relatively short period (the blink of an eye, in the evolutionary timescale) of the Cambrian explosion.

Better Red Than Dead: The Evolution of Warning Coloration

The lionfish (*Pterois miles*) is a majestic, mesmerizing fish. Watching it drifting slowly through the water in its native Indo-Pacific habitats, seemingly unaware of its natural splendor, is beautiful. Its body is brightly colored with bold maroon, brown and white zebra-like stripes.

The lionfish is a venomous species with 18 toxic spines that run along the length of its body. It is part of the Scorpaenidae family and related to the scorpionfish; however, unlike scorpionfish, lionfish are not camouflaged. On the contrary, they are highly conspicuous, especially against the blue water background, alerting any prospective predators to their toxicity.

Lionfish are an example of warning coloration that has presented an evolutionary challenge for the scientific community since Darwin's time. In 1866, Alfred Russel Wallace developed the term warning colors as a response to Darwin's

142 Homologous structures are similar structures that evolved from a common ancestor (for example, flippers and hands).

query regarding the bright colors of specific species of caterpillars. Darwin thought that males' bright colors could have evolved because of female aesthetic preference in sexual selection. However, caterpillars do not mate at this stage of their life, but only after they metamorphose into butterflies. In response, Wallace proposed that bright colors advertise the caterpillar's unpalatability to prospective predators.

The primary challenge that begs for an explanation is how warning coloration (aposematism) evolved. Why advertise a warning rather than blend in with the seabed or adapt other cryptic behaviors? Signals used by conspicuous prey increase distinctiveness and might increase the probability of attack from inexperienced potential predators. It is essential to bear in mind that when warning coloration first evolved, all predators were inexperienced, especially considering the aposematic prey's small population. Thus, the evolution of animal warning coloration is considered a paradox, mainly because it is assumed that the first conspicuous animal would be at a selective disadvantage and would eventually become extinct. In a way, this question resembles the discussion about the evolution of cleaning stations in Chapter 4. In both cases, when they first evolved, the animals put themselves at high predation risk, which raises the question of how such a trait could be genetically imprinted.

The underlying evolutionary principle behind warning coloration is the fast learning curve of predators. If the frequency of the appearance of conspicuous color patterns is too low, it might not facilitate the predator aversion learning curve. In other words, such signals only work once predators have learned to associate the conspicuous color with the unprofitability of the prey.

In general, conspicuous coloration can evolve, from a cryptic starting point, through three major possible avenues. The first avenue's sequence is that the prey becomes unprofitable first and only then evolves a conspicuous signal to publish its unprofitability. This scenario is considered the most likely evolutionary avenue since toxic defense production can be achieved in many ways (for example, by storing toxic materials or through symbiosis with bacteria). When an effective defense mechanism deters predators, the evolution of conspicuous coloration is likely since predators will kill less conspicuous defended prey by mistake. In the second avenue, the conspicuous signal appears first, followed by the prey becoming unprofitable. This route is considered less likely since the conspicuous prey's survivability, which is undefended, is slim unless the animal evolves Batesian mimicry, which means that a harmless species has evolved to imitate the warning signals of a toxic species. The last avenue is the simultaneous evolution of both conspicuous coloration and unprofitability. This alternative is also unlikely since, at the normal mutation rate, the chance of simultaneous mutations is very low.

Another hypothesis that might resolve this perplexing phenomenon is based more on the predator's foraging behavior than on the behavior of the conspicuous animal. This hypothesis assumes that warning colorations are not as risky as initially thought. It seems that novel prey shapes are not always at a selective disadvantage, since many predators have conservative and specific food preferences. Aviary studies demonstrate the importance of foraging preferences and the conservative decision that enables the fixation of a new novel shape. Thus, conspicuous aposematic coloration could evolve readily and repeatedly out of predators' conservative foraging decisions.

Warning coloration is also related to the handicap principle that was articulated in the mid-1970s by Israeli evolutionary biologist Amotz Zahavi (1928–2017), who argued that "in order to be reliable, signals have to be costly." For example, surviving with some functional disability signals superior fitness to potential mating partners. In other words, since animals have the evolutionary motivation to deceive each other, a signal will only be honest and reliable if it is costly for the signaler and therefore unattainable for low-quality individuals. Warning coloration reliability was substantiated by research that found that toxicity in nudibranchs and other sea slugs is positively correlated with conspicuousness, which gives one explanation for the perpetuation of warning coloration. On the other hand, the handicap principle does not predict Batesian mimicry, because conspicuousness is supposed to guarantee the honesty of a warning signal, and Batesian mimicry is, by definition, "not reliable."

This discussion addresses the selective force from the predator's perspective. However, many other biotic and abiotic factors also take part in the process of color fixation, and act antagonistically or synergistically with predator selection. An example is the trade-off between the positive effect of melanin (which traps heat) and the negative impact of increased predation risk.[143] An increase in melanin means a weaker and less effective aposematic signal, since it may limit the number of other pigments important in the warning signal. Parasite load is another example of a factor that influences color brightness. Empirical observations indicate that an increase in parasite load results in duller fish coloration.

In sum, the discussion regarding the evolution of warning coloration is a subset of the overall discussion about animal coloration. Ultimately, no single theory fully explains warning coloration since multidimensional aspects such as conventional signaling theory, the handicap principle, disruptive coloration, game theory modeling, and biotic and abiotic factors must all be considered.

143 Melanin is an effective absorber of light and a thermoregulator; it can dissipate over 99.9 percent of absorbed UV radiation. It also has a positive impact on the immune system.

‹ Spiny devilfish (*Inimicus didactylus*). Lembeh Strait, Indonesia.

‹ Yellowmouth moray (*Gymnothorax flavimarginatus*). Eilat, Israel.

"Geometry has two great treasures; one is the Theorem of Pythagoras; the other, the division of a line into extreme and mean ratio. The first we may compare to a measure of gold; the second we may name a precious jewel."

Attributed to Johannes Kepler (1571–1630)

CHAPTER 9

Mathematical Beauty in Coral Reefs

« Sea stars are characterized by pentagonal symmetry, which is associated with the Golden Ratio. Puerto Galera, Philippines.

As part of his immense contribution to Western philosophy, Greek philosopher Plato presented the concept of the sensible, observable world as an imperfect image of an ideal realm related to constant, abstract objects. This distinction has accompanied Western philosophy ever since, and is reflected in the thoughts of many contemporary scholars. Plato's philosophical principles are central to Western thought, as expressed by English mathematician and philosopher Alfred North Whitehead (1861–1947): "The safest general characterization of the European philosophical tradition is that it consists of a series of footnotes to Plato." Plato and his disciples reflected the natural human ambition to understand the universe's eternal order and beauty through mathematical patterns. Moreover, Plato was one of ancient Greece's most significant mathematics patrons, having founded the Academy of Athens in 387 BCE; tradition says that the sentence "Let no-one ignorant of geometry enter here," was engraved at the door.

The connection between beauty and numbers was also found in Roman thought. Roman architect Marcus Vitruvius Pollio (70–25 BCE) is famous for his study of human body proportions, as drawn by Leonardo da Vinci in the "Vitruvian Man." In his book about architecture he wrote: "in the human body there is a kind of symmetrical harmony between forearm, foot, palm, finger, and other small parts; and so it is with perfect buildings."[144]

More than 1,400 years later, Galileo Galilei (1564–1642) continued the Greek–Roman tradition, stating that our universe is a "grand book" written in the language of mathematics. A century later, French scholar Pierre-Simon Marquis de Laplace

144 Vitruvius, P. (1901). Vitruvius: *The Ten Books on Architecture*. Forgotten Books.

> The School of Athens (Raphael, 1509–1511). The two main figures in the center are Plato (left) and Aristotle (right). Plato's gesture toward the sky is thought to indicate his theory of eternal forms; the position of Aristotle's hand is thought to indicate his belief that knowledge comes from experience. Plato is probably a portrait of Leonardo da Vinci. (For full credit information, see Figure credits.)

(1749–1827) expressed the idea of scientific determinism, arguing that if we knew the positions and velocity of all the particles in the universe at a specific point in time, then we could calculate and know their behavior at any other point in time, in the past or future. Laplace's vision implies that we can predict the future, at least in principle. This idea was a central tenet of science until the last century.

Since then, two significant developments have shown that Laplace's view is not applicable, and that mere determinism is far from reality. The first development was quantum mechanics, asserting inherent uncertainty regarding the simultaneous measurement of space and time (Heisenberg's uncertainty principle).[145] In the spirit of Laplace, scholars like Albert Einstein (1879–1955), known for saying "God does not play dice with the universe," criticized Heisenberg's uncertainty principle. Einstein argued that there are "hidden variables," which have not yet been identified, and might explain quantum mechanics' statistical phenomena. The second development is a set of theories broadly named complex dynamic systems, contending that due to the sensitivity and complexity of non-linear systems, we are unable to predict the

145 The principle is named after Werner Karl Heisenberg (1901–1976), a German theoretical physicist and Nobel Prize winner who was one of the founders of quantum mechanics.

future. An example of this is the "Butterfly Effect" taken from chaos theory. The latter development did not exclude determinism, but did assert that determinism does not necessarily mean predictability.

Biology and ecology are sciences of systems and are therefore no exception. There are significant challenges in predicting the trajectories and behavior of ecological systems (such as coral reefs), to say the least. However, mathematical models can provide us with significant insights and the ability to identify regularities in biological systems. This contribution will likely be augmented by advances in computer science modeling and an increase in computing power.

Mathematics is not as abstract as many people think. It has representations in all physical systems, from the quantum level to the cosmos. It serves as a fundamental method in understanding the universe and reflects the inherent human pattern-like thinking and the desire to explore nature's underlying laws. Even today, the underlying thoughts behind many scientific endeavors have some similarities with Plato's ideas.

The Divine Proportion

Ancient Egyptians were among the first cultures to base their model of human aesthetics on a canon that included a set of fixed proportions between parts of the human body. More than two millennia later, Greek mathematicians and philosophers contributed to the formulation of divine mathematical laws, the numbers Pi ($\pi = 3.14...$), and the golden ratio phi ($\phi = 1.618...$), named after Phidias, the great Greek sculptor.[146] Since then, the ratio and Pi have accompanied aesthetic ideals for more than 2,500 years.

During the Renaissance, there was a revival of major concepts contemplated by Greek philosophers and mathematical proportions were re-sanctified. These concepts were central to the work of many Renaissance artists, including Piero Della Francesca (1416–1492) and Leonardo da Vinci (1452–1519), the latter of whom was probably the first to coin the term *Sectio Aurea* or golden section. Franciscan monk and mathematician Leonardo Fibonacci (1170–1250) contemplated the Fibonacci sequence,[147] a variation of the golden ratio, considered an important mathematical model to describe many of nature's patterns. The earliest known treatise about the

146 Phi ($\phi = 1.618$) is the positive solution of the simple quadratic equation, $x^2 - x - 1 = 0$.

147 The Fibonacci sequence is a sequence in which each number is the sum of the two preceding ones (0, 1, 1, 2, 3, 5, 8, 13, 21, 34... FN + 2 = FN + 1 + FN). The ratio between two consecutive Fibonacci numbers converges to the golden ratio (1.618).

golden ratio was published during the Renaissance by Luca Pacioli (1445–1517) and named *Divina Proportione*. Pacioli's book, dedicated to a comprehensive description of the immense variations of the golden section, can serve as an example of its centrality in Renaissance art.

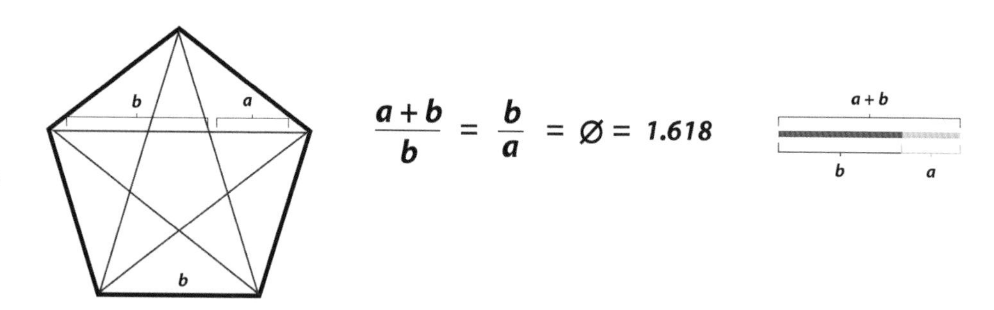

> The golden ratio is defined as the ideal proportion between two parts of a line, where the ratio between the long line (b) to the short line (a) equals the ratio between the entire line (a+b) to the long line (b). In other words: the whole to the longer equals the longer to the shorter. The pentagram is a variation of the golden ratio, and both a magical and holy symbol.

The journey towards the ultimate definition of aesthetics, based on the golden ratio, continues to play a significant role in modern thought. Swiss architect Charles-Édouard Jeanneret (1887–1965), known as Le Corbusier, adopted the golden ratio as a central motif in his work. In 1954, he published The Modulor: A Harmonious Measure to the Human Scale Universally Applicable to Architecture and Mechanics, a book in which he set a series of measurements to define the preferred and optimal proportions of the parts of a building based on a human scale and the golden ratio.

The golden ratio has derivatives and manifestations in many life phenomena. To name just a few: geometrical forms (such as the golden triangle, rectangle and pentagon), the Fibonacci series, human body proportions, and the golden spiral, which is very abundant in nature. One variation of spirals is the logarithmic spiral; it can be found in galaxy patterns (such as our galaxy, the Milky Way), hurricanes, DNA's double-helix, the trajectory of a hawk approaching its prey, and more. The prevalence of logarithmic patterns in nature is probably why humans often attribute sacred qualities to them. It is no wonder that the ancient sect of the Pythagorean, who believed that numbers rule the universe, were committed to an oath of silence regarding the mysterious rules that dominate its pattern.

Coral reefs and marine creatures are no exception, and provide us with the opportunity to observe some of the underlying mathematical proportions that govern the world. We can observe the golden ratio in many species of marine organisms, especially mollusks. Examples are the spirals of Christmas tree worms, nautiluses, seashells and more. Moreover, many marine animals, amongst them some 1,500 species of starfish, are characterized by pentagonal symmetry, which is closely related to the golden ratio.

Numerous seashells manifest variations of a geometrical logarithmic spiral in which the distance between the sections increases in geometrical sequence. In shells, its appearance can be understood as being created by the mineral aggregation of two combined vectors: one in a radial direction and the other in a tangential direction. The magnificent, logarithmic spiral growth which we see in shells is constructed by new material that is continually added at the open end of the shell. If the proportion of the magnitude of these two vectors is constant during the shell's entire growth process, the vector's angle remains the same, and a logarithmic spiral appears. It is important to note that while not all the logarithmic spiral ratios represent the golden ratio, they all seem to demonstrate beauty and perfection, along with other ratios.

Golden spiral in a nautilus

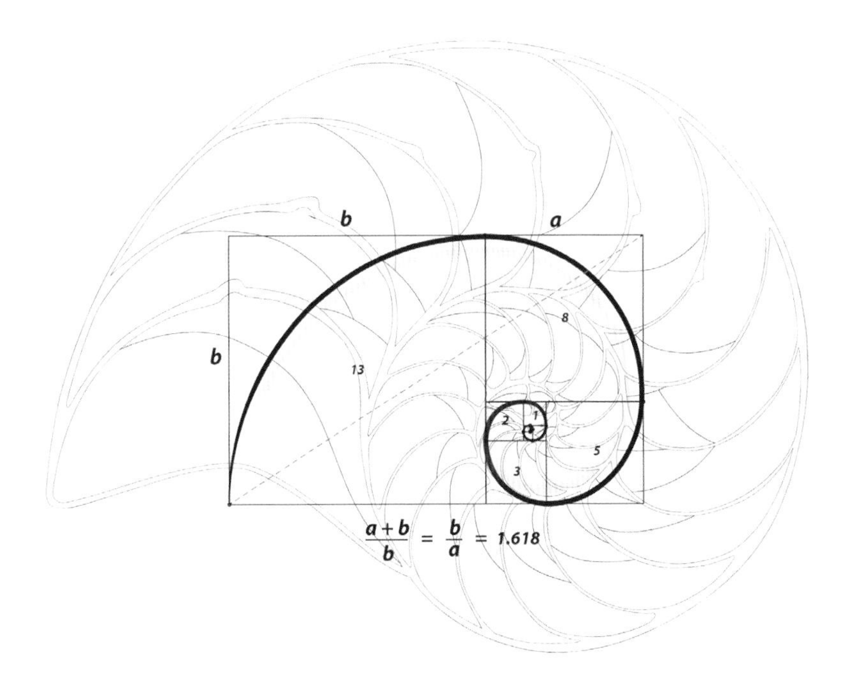

From a biological perspective, spirals represent an efficient spontaneous growth pattern. They are also energetically cost-effective by maintaining the same shape through each successive turn of the spiral. As pointed out by astrophysicist Mario Livio (1945–), self-similarity is the essential feature of a logarithmic spiral. This feature

is a prerequisite for many growth phenomena in nature, since it enables a creature to grow without losing its basic structure. For example, as a mollusk grows inside its chambered nautilus shell, it requires an increasingly larger space. Subsequently, each increment in the shell's length is accompanied by a proportional increase in its radius. The shape remains similar and saves adjustments to its body symmetry.

For thousands of years, the golden ratio has played a significant role in the history of thought. The basic idea that mathematical rules and proportions represent beauty and perfection, and at the same time induce comfort and harmony, remains vital and inspiring. In its essence, this means there is order in the mysterious universe that surrounds us, and humankind has managed to explore some of its secrets.

« Logarithmic spiral in a Christmas tree worm. Verde Island, Philippines.

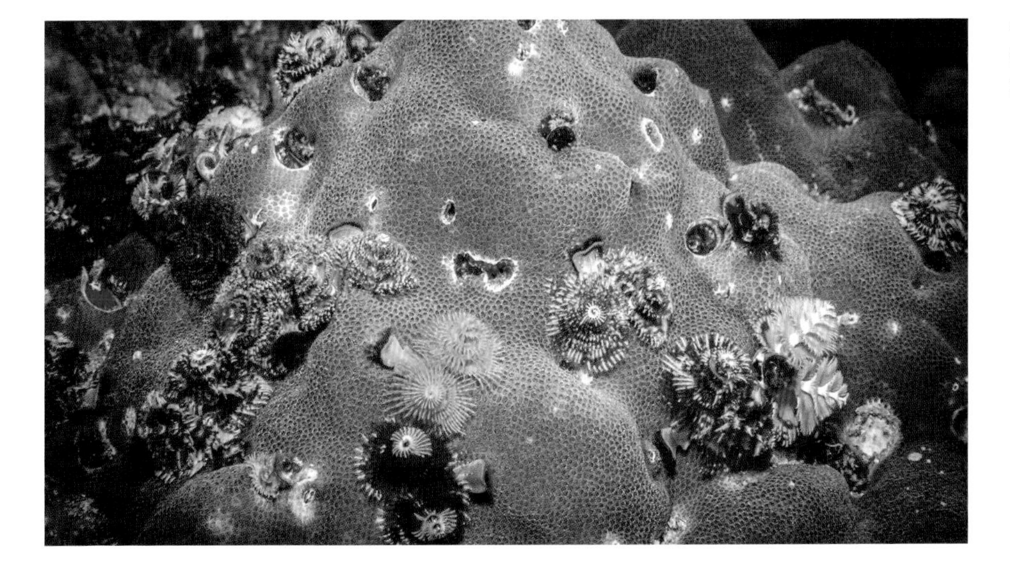

‹ A garden of Christmas tree worms. Verde Island, Philippines.

Symmetry and the Evolution of Flatfishes

Symmetry is ubiquitous, and an essential part of most patterns in nature. It is impossible to exaggerate the importance of symmetry, since humans are universally obsessed with symmetrical patterns, as reflected in art, storytelling, architecture and science. For example, in mechanical and quantum physics, every conservation law (including the conservation of energy, linear/angular momentum and electrical charge) is derived from various rules of symmetry (Noether's first theorem).[148]

148 A conservation law is a physically measurable quantity that does not change as a result of any physical process (in a closed system, the total energy is constant before and after any process occurs). Noether's theorem is named after German mathematician Emmy Noether (1882–1935).

Noether's groundbreaking theorem embeds the direct relationships between nature's dynamic forces and motion, and the abstract world of continuous symmetry (mathematics). In other words, Noether's theorem tells us how the symmetries of the universe are manifested through physical laws in nature.

Symmetry is also the only trait to which we can attribute an optimal value. Most creatures in nature aim to achieve perfect symmetry. In humans, it is in the range of 1 percent deviation of external symmetry. Symmetry is associated with fitness, and serves as an index of fertility and genetic quality. Moreover, numerous biological studies indicate that animals prefer symmetrical mating partners. As an example, research conducted in Jamaica found that knee symmetry alone predicts an athlete's sprinting speed and the best of the best runners.

In nature, a different form of symmetry can be seen in plants and animals, and serves biologists who describe and classify animals. Some primordial animals, for example, sponges, do not have any type of symmetry. These animals, usually simple multicellular ones, are characterized by simple body plans, no internal organs, and the lack of a centralized nerve system. Other, more complex animals are categorized by various types of radial or bidirectional symmetries.

Radial and bi-radial symmetries have body parts arranged around a central point. Animals with radial symmetry such as sea urchins, corals and sea anemones have many lines of symmetry. Most of these animals live in water and do not move quickly, but have the advantage of being able to sense food and danger from any direction.

Bilateral symmetry, the most common type of symmetry (99 percent of species), has one line that divides the animal into two halves. It has several advantages over radial symmetry, included directed locomotion, which requires a distinct front end (head) with sensory organs required to gather and process information in a particular direction. Directed locomotion enables organ specialization and allows creatures to create three-dimensional images and use ears to detect the target direction. Recent phylogenetic research proposes that bilateral symmetry may have evolved out of the need to improve the efficiency of internal circulation, and that it pre-dates directed locomotion. One example supporting this hypothesis is bilateral symmetry in some species of cnidarians, which are mostly characterized by radial symmetry (for example, the bilateral sessile sea pen).

Flatfishes such as flounder, sole, halibut, turbot, plaice and tonguefish are among the most asymmetrical vertebrates ever to live on Earth. They are characterized by having both eyes located on the same side of the head, an enigma that set an explanatory challenge for evolutionary biologists.

Although flatfish larvae look like ordinary symmetrical fish larvae, they are genetically programmed to become asymmetrical creatures. As they begin their way into adulthood, they change from being symmetrical little fish into anatomically asymmetrical creatures. They lose one side of their face, the bones in their skull bend, and one eye moves to the other side of their head. Eventually, flatfishes sink to the sandy bottom of the sea, settling on the blind side of their face, with both eyes located on one side, a distorted mouth, and one fin flattened towards the sandy bottom.

In fact, they undergo additional post-embryonic asymmetric remodeling during metamorphosis, which is completely different from the bilateral symmetry that characterizes the vast majority of vertebrates. From an early stage, the flatfish's inner ears, which together with its eyes are responsible for upright swimming orientation, begin to override the eyes until they become the dominant organ responsible for their sense of direction.

One open question pertaining to this sequence of development is whether the shift in the eye position (as a result of mutation) preceded the flatfishes' bottom-dweller behavior, or vice versa. Researchers believe that the anatomical makeover of the flatfish followed a change in its behavior because the bottom of the sea was probably a place with relatively little competition.[149]

The bizarre structure of flatfishes serves creationists and opponents of Darwinism to argue for the implausibility of evolution theory. If flatfishes evolved from symmetrical ancestors, how did they survive the transition period of facial reconstruction? Creationists preach for divine intervention, and ask how the perfect fit between an organism's form and function could possibly be explained with science.

This question is at the heart of the longstanding debate between evolutionists and creationists. Creationists use the watchmaker analogy to ridicule skepticism about divine intervention. The analogy was presented by English clergyman and philosopher William Paley (1743–1805), who lived before the publication of the evolution theory. Paley argued that a godly design is required to create the complex structures of organisms. He claimed that if you find a pocket watch on the ground, you should assume that it was constructed by a watchmaker, and not created naturally without intelligent design. In a teleological manner,[150] Paley made the analogy between God and a watchmaker, where the watchmaker is to the watch as God is to the universe.

149 https://www.pbs.org/wgbh/nova/article/flatfish-evolution/ (retrieved December 2, 2022).

150 Teleology is an explanation for something as a function of its end, purpose or goal.

Following the advent of evolution theory, we know that it would be a mistake to try and explain the asymmetry of the flatfish in a teleological manner. Evolution is based on tinkering, the construction or creation of something from a diverse range of available materials. After all, flatfish demonstrate that the most critical success factor is survivability and not symmetry...

^ Flatfish such as this flounder are well-camouflaged, excellent predators, with 360° vision and very fast gulping capability. While octopuses control their pigment cells via the nervous system, the color change in flatfishes is controlled by the release of chemicals into the blood. Though it is a slower process, it provides flatfish with excellent disguise capabilities. Lembeh Strait, Indonesia.

The Hyperbolic Universe

In his book *The Wonderful Century: Its Successes and Its Failures* (1898), co-founder of the evolution theory Alfred Russel Wallace wrote:

> *Then, going backward, we can find nothing of the first rank except Euclid's wonderful system of Geometry, derived from earlier Greek and Egyptian sources, and perhaps the most remarkable mental product of the earliest civilizations; to which we may add the introduction of Arabic numerals, and the use of the Alphabet.*

Humans construct the world mainly with straight lines, which are based on Euclidean geometry. Most artifacts, houses, skyscrapers and streets are structured on this type of geometry. Nature, on the other hand, is different.

Euclidean geometry, founded by Greek mathematician Euclid of Alexandria (323–283 BCE), fundamentally includes five axioms. Those axioms were assumed to hold the ultimate truth and serve as the basis to prove all other propositions. The fifth axiom of Euclidean geometry is that parallel lines (believed to be infinite) never meet, and that the angles of any triangle always add up to 180 degrees. Unlike flat Euclidian space, spherical and saddle-like spaces are described by different geometry, and its axioms differ from those of Euclidean geometry. On a curved saddle shape, the sum of a triangle's angles is always less than 180 degrees. This type of geometry, called hyperbolic geometry, was independently articulated at the beginning of the nineteenth century by a few scholars, including German mathematician and physicist Carl Friedrich Gauss (1777–1855) and Russian mathematician Nikolai Lobachevsky (1792–1856). The only difference between Euclidean and hyperbolic or spherical geometries is in the axiom of the parallel line. Many of the geometric rules taught in school do not apply to hyperbolic surfaces. For example, the circumference and the circle area are smaller than $2\pi r$ and πr^2, respectively.

Euclidean versus **Non-Euclidean Geometry**

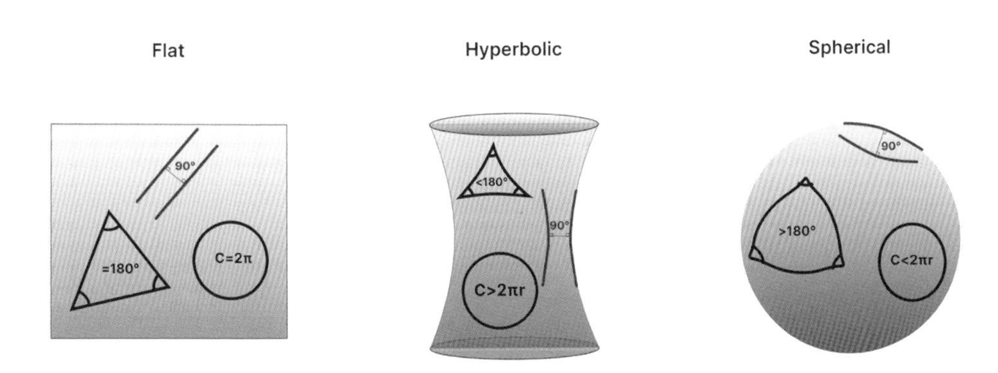

Non-Euclidian geometry challenged the view that all physical space can be described by Euclidean geometries, a concept that had been perceived as an absolute and eternal truth until Gauss. Non-Euclidian geometry raised questions about the relationships between science, mathematics and observations. At a later stage, this discovery paved the way for Einstein's theory of general relativity, where gravity is often visualized as curved space.

> Hyperbolic plane of a yellow scroll coral (*Turbinaria reniformis*), a type of stony coral. Eilat, Israel.

˅ Hyperbolic plane of a dark margin glossodoris nudibranch. Lembeh Strait, Indonesia.

Based on general relativity, mass causes space to curve, so the entire mass (strength of gravitation) of the universe is presumed to determine the universe's structure. One hypothesis proposes that this structure is hyperbolic, which means there is an insufficient mass to slow down and cease its expansion. In other words, the universe will expand forever. Other excellent examples of hyperbolic planes are some of the works by Dutch artist M. C. Escher (1898–1972).[151]

Like many terrestrial organisms, the shapes of numerous marine creatures, including sea slugs, sponges and corals, are characterized by frilly, non-Euclidian patterns. In coral reefs, many nudibranchs, worms, seaweeds, and corals demonstrate hyperbolic structures.[152] These forms are recognizable by the frills and crenellated shapes of the body and edges. Evolutionarily, wherever there is an advantage to maximizing the surface area of a given radius, such as for filter-feeding animals or the absorption of light by corals, hyperbolic shapes are more effective than flat curves.

Like the golden section, hyperbolic planes are universal phenomena, from sea slugs to the universe's entire structure. This remarkable multi-scale universality and unity are expressed in the words of English poet and painter William Blake (1757–1827), who wrote in his poem "Auguries of Innocence":

To see a World in a Grain of Sand
And a Heaven in a wild flower
Hold Infinity in the palm of your hand
And Eternity in an hour...

"Endless forms most beautiful" – Turing Patterns

From the beautiful colorful patterns of mandarinfish, emperor angelfish and parrotfish to the bright dark-edged blue dots of the coral grouper, periodic patterns abound in nature. The evolutionary rationale for these patterns was discussed in Chapter 3, in the context of their adaptive role in signaling and camouflage by disruptive coloration. Nevertheless, another central question regarding this topic is how can such patterns, so complex and diverse, evolve during an organism's development? The problem arises because most animals have no related structures under the skin. Therefore, patterns must be created by the skin cells, without the support of an underlying structure.

151 https://ww2.amstat.org/mam/03/essay1.html (retrieved November 16, 2022).

152 https://aeon.co/essays/theres-more-maths-in-slugs-and-corals-than-we-can-think-of (retrieved November 16, 2022).

It makes sense to assume, therefore, that there is an underlying mechanism that combines genetics with physical and chemical rules that repeatedly create those patterns.

Alan Turing (1912–1954) was an English computer scientist and philosopher who was also the father of theoretical computer science and artificial intelligence. During the Second World War, Turing was extremely instrumental in the cryptanalysis that led to breaking the code of the German cipher machine Enigma, a decisive factor in the Allies' victory. He later turned his attention to mathematical biology, publishing a groundbreaking paper entitled, "The Chemical Basis of Morphogenesis," discussed further below.

Tragically, Turing was persecuted by British authorities for being homosexual, accusations which are inconceivable today. He was sentenced to chemical castration and died shortly thereafter, probably due to suicide. It was not until 2009, more than 50 years after his death, that British authorities apologized for the appalling way they had treated him. Today, he is celebrated as a hero and a trailblazer.

φ φ φ

Turing was the first to contemplate the idea that random activity at the molecular level could result in large scale order. In "The Chemical Basis of Morphogenesis," he presented a mathematical model that describes the behavior of a system in which two substances interact with each other and diffuse at different rates.[153] This single pattern-formation mechanism, termed reaction–diffusion,[154] simulates many of the patterns observed in nature, including the colorful designs that decorate many reef creatures.

Turing described two chemicals, one that acts as a reactor and another that inhibits the reaction from happening. When these two chemicals spread and diffuse through a group of cells, each at a different pace, they spontaneously self-organize into spots, stripes, rings, swirls, etc. Turing proved mathematically that such a system could form a regular pattern that emerges from an initially homogeneous state. The complex patterns we observe are not necessarily perfectly symmetrical and without accurate regularity. These emergent patterns are in line with the behavior of non-linear equations, the same category of equations that describe chaos theory.

---

153 Diffusion is the movement of a substance (such as atoms and ions) from an area of high concentration to an area of low concentration.

154 A popular explanation of the reaction–diffusion process can be found here: https://youtu.be/alH3yc6tX98 (retrieved March 12, 2021).

Numerous mathematical models simulate pattern formation, but Turing's model presents it simply and elegantly. Suppose we follow the philosophical principle called "Occam's razor," which suggests that we favor the simplest solution in the face of several competing solutions for a single problem. In such a case, Turing's model is an excellent candidate.

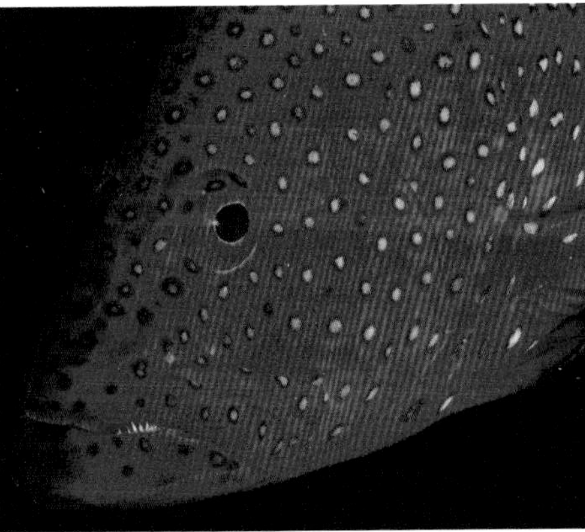

< Different skin patterns on a Red Sea racoon butterflyfish (far left) and a grouper (left). The exact pattern depends on the shape of the skin and the diffusion speed of the activator and inhibitor. Eilat, Israel.

Turing's model is an example of the power of mathematical models to describe biological phenomena. Since around the turn of the millennium, researchers have demonstrated the accuracy of Turing's theory and that it is plausible to assume that such a mechanism exists in the bodies of animals. An example of this can be found in Hans Meinhard's 2009 book The Algorithmic Beauty of Seashells, in which Meinhardt demonstrates the applicability of Turing's equations to seashell patterns. Another example is research by Japanese scientists Shigeru Kondo and Rihito Asai, who photographed the patterns on one species of angelfish and found that predictions based on Turing's model were correct, and that the patterns move at a predictable pace.

The challenge now for developmental biologists and geneticists is to determine and demonstrate, on a living organism, the precise molecular changes and evolutionary processes responsible for this incredible diversity of color patterns.

Coral Reefs as Complex Systems

On December 26, 2004, an undersea earthquake with an exceptional magnitude of 9.1 caused a deadly tsunami that struck coasts across the Indian Ocean and the Andaman Sea. For the next seven hours, immense ocean waves up to 30 meters high devastated coastal communities in 14 countries, killing an estimated 230,000 people.

While many Thai fishing crews were caught by the waves and died, the Moken sea nomads survived. Though waves completely destroyed their 200-person village, only one person perished. The Moken sensed the arrival of the tsunami and searched for higher ground or headed to the deep sea. They probably recognized the tsunami's arrival due to the strength of the first wave, the dramatic retreat of the sea, or the sight of the next waves to come. Previous tsunami event signs had been passed on orally for generations, from parent to child, and were stored in the Moken's collective memory. For them, this was not a "black swan" event.[155] The animist Moken believe that a *laboon*, "a wave that consumes people," was sent by ancestors' spirits to wipe out the world's evils, demolishing everything in its path, before all is reborn.

In February 2007, two years after the tsunami, I traveled to the Similan Islands archipelago in Thailand to dive. During my diving sessions, I saw large reef areas that had been completely flattened by the tsunami.[156] However, it was intuitively clear to me that the dynamics in those reefs indicated a recovery process. Two years later, in April 2009, I traveled again to the Similan Islands and was impressed by the speed of the reef's recovery and the species diversity. At that time, I was especially intrigued by the question of how reef systems recover without a master plan (a blueprint) and achieve a system equilibrium, possibly different from the previous one, composed of an infinite number of interacting organisms.

φ φ φ

The process of reef recovery from such an acute natural disaster is called secondary ecological succession. The term also refers to a community's normal development from the early settlement stages to its climax, when equilibrium between species and environment occurs. This process is shared by some complex natural systems and characterizes the recovery of many natural habitats following either natural or human-caused disasters (for example, volcanic eruptions, earthquakes, wildfires and dynamite blasting). This recovery process indicates the self-organization of coral reefs, which means some degree of order can be found in the system.

---

155 A black swan is an unpredictable, rare event beyond what can be expected of a situation and with potentially catastrophic consequences. The term is based on an ancient tradition that assumed black swans did not exist. This belief was refuted following the discovery of black swans in Australia in 1697.

156 It is estimated that 13 percent of coral reefs in that area suffered severe (more than 50 percent) damage from the tsunami (https://wwf.panda.org/wwf_news/?18650/post-tsunami-coral-reef-assessment-done-in-thailand) (retrieved March 12, 2021).

The rate and scope of a coral reef's recovery are highly dependent upon the magnitude of the disturbance with regard to factors such as pre-disturbance conditions, competition, predation, dispersal and legacies, as well as major environmental factors such as light, temperature and salinity. The recovery process is also dependent on the nature of the disturbance, whether it is acute (severe and sudden) or chronic (moderate and long-term). These factors and others will dictate the pace of recovery and the new reef's structure, either similar to the previous one or different in varying degrees.

Coral reefs have been seen to recover reasonably well from some acute perturbations; however, in many cases they do not recover well from chronic disturbances of either natural or human origin, or gradual declines in environmental conditions. Unlike the acute tsunami of 2004, the problem with global warming is chronic, and may well cause irreversible damage or change coral reef systems as we know them today.

Mechanistic versus holistic worldview

A natural conclusion from the above is that understanding the reef system's ecological succession and other characteristics requires the adoption of a holistic perspective. In fact, the holistic, organic worldview is not a new human idea. Until the sixteenth century, for more than two millennia, prevailing thought rested on the religious and Aristotelian ideas that had seen the world as organic, living and spiritual. During the sixteenth century, a new paradigm was formulated: the mechanistic worldview. This paradigm perceived the world and nature as machine-like and made up of parts (a bottom-up approach). The mechanistic worldview started to be replaced during the twentieth century by a paradigm called systems thinking, somewhat resembling the ancient organic viewpoint. Systems thinking perceives the world as holistic and better described by the metaphor of the world as a network. The underlying tension between these two different (though in my opinion, complementary) schools of thought are between the parts and the whole, between dichotomous and holistic thinking.[157]

An example of this re-emerging holistic thinking is Gestalt psychology, which dealt with the study of perception. The central tenet of Gestalt psychology is that organisms perceive their environment in holistic perceptual patterns and not by individual components. One of the founders of Gestalt psychology, Austrian philosopher Christian

157 Dichotomous thinking, one of the foundations of Western thought, has its roots in the Greek Platonic distinction between the ideal world and the real one, the intellect and emotion. These dualities are at the heart of the Judeo-Christian worldview, where the main dualities are between spirit and flesh, humanity and nature, sin and redemption.

von Ehrenfels (1859–1932), coined the phrase, "The whole is more than the sum of its parts," which later became the slogan of systems-thinking leaders.

Ecology has undergone similar changes. Historically, ecological science studied the interaction between organisms and their environment while distinguishing between the two entities. Today, ecology has changed its perspective, extending its focus to the identity between objects and their environment.[158] Based on this school of thought, we cannot separate between sea anemone and clownfish, or between coral and algae, and must refer to them as single entities. The terms superorganism and holobiont, discussed in Chapters 2 and 5, are part of this emerging holistic view.

The reef as a complex dynamic system

The emergence of integrative theories and powerful mathematical models such as cybernetics and complex dynamic systems theory are outcomes that reflect a conceptual change.[159] In recent years, complex systems theory has been sharpened from an abstract concept into a series of tools that can be used to describe concrete living systems. These theories could provide insights that improve our understanding of complicated interacting ecological systems such as coral reefs. Complex systems theory could address topics such as modeling ecological succession, reef system recovery, and resistance in the face of acute, chronic or temporary perturbations.[160]

There are many definitions for the term "complex systems." Perhaps the simplest one was articulated by the editors of *Science* magazine, who defined it as "one whose properties are not fully explained by an understanding of its component parts."[161] Complex systems are difficult to predict and are characterized by sensitivity to initial conditions and external influences, sometimes resulting in chaotic behavior.

The unique and large number of interdependent interactions between reef creatures is among the primary reasons for the coral reef system's resilience for millions of years. Features such as redundancy and feedback are characteristics that

---

158 https://aeon.co/essays/not-just-the-lotus-blossom-buddhism-and-ecology-partner-up (retrieved March 12, 2021).

159 Cybernetics deals with the control, communication, regulation and constraints of animals and machines. The concept was coined by Norbert Wiener (1894–1964), author of *Cybernetics: Or Control and Communication in the Animal and the Machine* (1961).

160 Resistance is the degree to which a system can withstand stress, such as disturbance in its internal or external environment, without losing its functionality.

161 This definition is from "Beyond reductionism," published in *Science* in 1999. A more formal definition, from "Incorporating complexity in ecosystem modelling," in *Complexity International*, is: "a network of many components whose aggregate behavior is both due to and gives rise to, multiple-scale structural and dynamical patterns which are not inferable from a system description that spans only a narrow window of resolution."

< The white corals in the photo were bleached due to an extremely low tide that exposed them to the air and direct sunlight, not as a result of global warming. Some are expected to recover within a period of weeks. Eilat, Israel.

support coral reefs in coping with acute perturbations such as tsunamis and also, to a limited extent, with coral bleaching.

Coral bleaching is a phenomenon in which the coral expels large quantities of algae, the symbiotic partner that live inside it, as a reaction to stress. Often, this stress is caused by chronic global warming. As a result of the expulsion, the coral loses its color and may become less fit. Since many corals are dependent on these algae for energy, they often die following bleaching. Some scientists believe corals are characterized by functional redundancy that can alleviate the damage caused by coral bleaching through a process in which corals replace their dominant existing symbiotic algae with other algae species, ones which are more tolerant to environmental changes. By replacing the algae, corals may be able to tolerate some stressful conditions, such as a rise in water temperature. If this is correct, though some scientists think it is not, bleaching may be considered an adaptive response that sacrifices short-term benefits for long-term advantages, achieving resistance and higher fitness.

One of the most abundant features in complex systems is feedback loops that lead to self-correction; these can be found in coral reefs, as well as many other biological systems. It seems that coral bleaching, which remains a poorly understood phenomenon from many aspects, includes an unusual regulating feedback loop mechanism.

In a study by Bollati and colleagues, for instance, it was found that in some instances, bleached corals become, within a few weeks, remarkably colorful (green, yellow or purple-blue) following exposure to temporary heat stress. That study suggests that an optical feedback loop regulates the relationships between the polyp and the symbiotic algae.

Following coral bleaching, the coral becomes white and is penetrated by blue light that was previously absorbed by the yellow-brownish algae. This blue light induces the expression of colorful pigments and as a result, the colorful coral reduces internal light stress and facilitates the recolonization of the algae's bleached tissue. The colorful coral then starts recovering and goes back to its original coloration due to the algae's re-colonization. The loop is closed by the coral's re-colonization by algae that reduces the influx of internal blue light, and switches off the expression of colorful pigments.

An additional example that shows a feedback loop in coral reefs is their function as a giant thermostat, regulating the temperature of their environment. In the Great Barrier Reef, for example, this regulation occurs along an immense area that is 2,300 kilometers long. When stressed by heat, the algae that live inside the corals produce compounds that make a gas, which creates an aerosol layer just above the water. This layer scatters the sunlight and creates tiny bright clouds that shade and cool the water around the reef. Research conducted in the Great Barrier Reef has shown that over the long term, the temperature of the water around the reef increased a little less than in other places. In other words, it was posited that, to a certain extent, coral reefs might be able to protect themselves from the stress caused by heat by emitting a gas that serves as a shield against solar radiation. Needless to say, the term "giant thermostat" might give the wrong impression. While this is remarkable, it is insufficient for coping with the acute and chronic impacts caused by humans and global warming.

Another feature of complex systems evident in reefs is the feedback loop in the correlation between the number of parrotfishes and their body size, and the extent of bleached coral in a given area. The interaction between corals and parrotfishes constitutes a negative feedback loop. When reefs are damaged, parrotfish numbers increase, as does the pace of their scraping corals and cleaning off the scunge. This scraping gives the coral a push towards recovery and replenishment. At the same time, the number of other fish species declines. Furthermore, individual parrotfish living near the bleached coral area were 20 percent larger than those living in a healthy reef. When the reef recovers, the opposite processes occur.[162]

162 https://www.upi.com/Science_News/2019/12/02/Parrotfish-thrive-in-the-wake-of-coral-bleaching/6711575304910/ (retrieved January 17, 2021).

< Parrotfish. Eilat, Israel.

< The teeth of the green humphead parrotfish are fused, enabling it to scrape corals and remove scunge, which pushes the coral towards recovery and replenishment. Similan Island, Thailand.

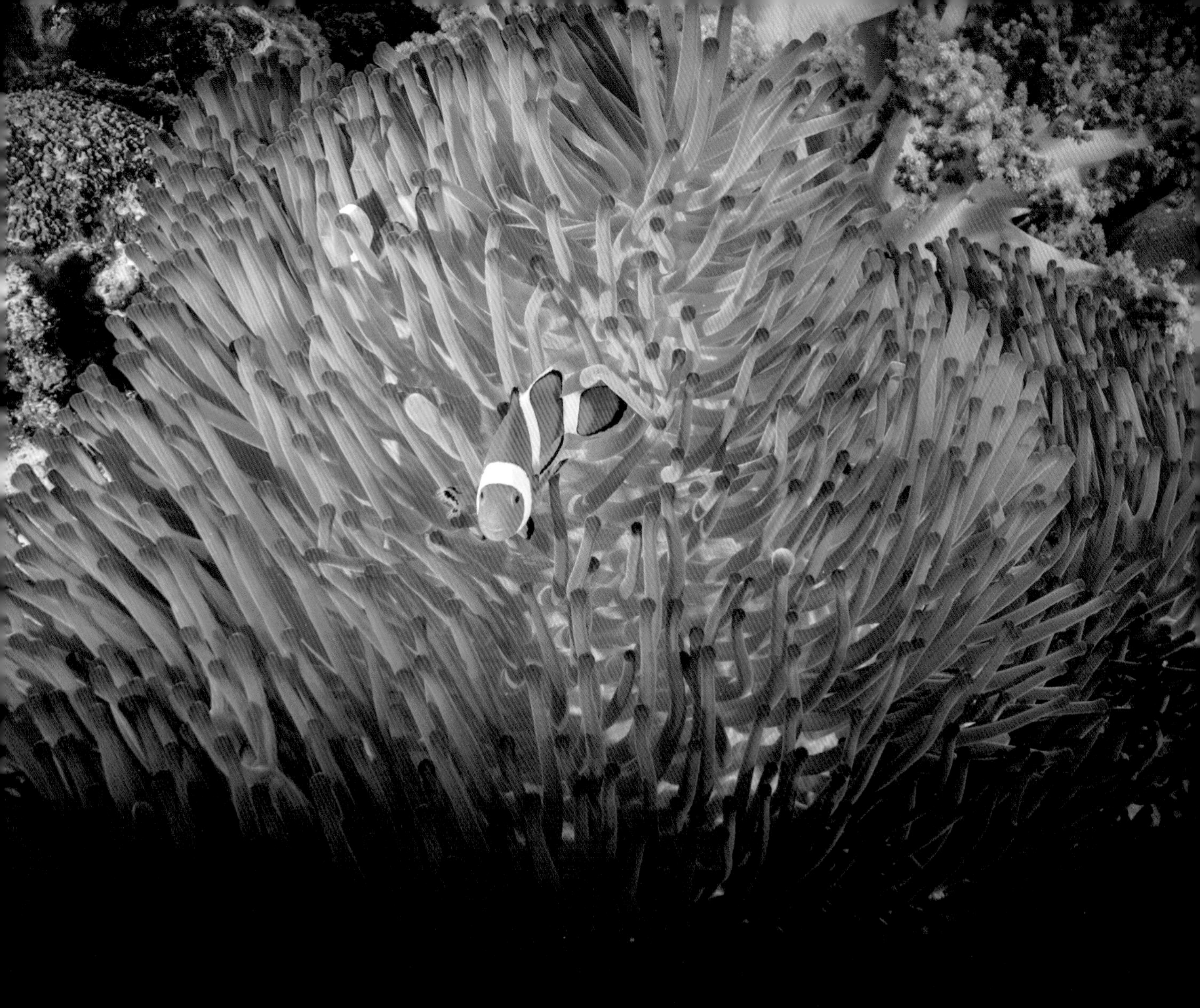

"We are caught in an inescapable network of mutuality, tied in a single garment of destiny. Whatever affects one directly, affects all indirectly."

Martin Luther King, *Gospel of Freedom: Martin Luther King, Jr.'s Letter from Birmingham Jail and the Struggle that Changed a Nation*, 1963

Epilogue

The intellectual journey into the world of coral reefs presented in this book is only a glimpse of this magnificent natural wonder. The problem is, this story does not have a happy ending. The coral reef ecosystem is in great danger, and it seems that future generations might not have the privilege of enjoying its scenes of natural wonder. Coral reefs face a fatal combination of acute and chronic impacts. Their well-known, longstanding resilience might be insufficient for coping with such a devastating combination, one that may irreversibly damage global marine biodiversity, as well as the welfare of the local communities that rely on them.

« Clownfish and sea anemones always live together. Does this make them a single evolutionary entity? Bunaken, Indonesia.

The phrase "so many words were said, so little was heard," has never been more relevant than it is in connection with global warming. The tremendous amount of information provided by both the scientific community and the media has been insufficient to change our behaviors on a scale that is significant enough to end coral reef deterioration.

Climate change can be considered a perfect problem. It is amorphous, obscurely defined, with no precise geographic location, no single cause, and no clear enemy. Furthermore, as part of the postmodern world we live in, there is a continuous decrease in the authority of public institutions and a corresponding increase in the ability of social media to provide platforms for charlatans to express their opinions and spread "alternative facts" that undermine the messages of the scientific community.

One of the reasons for our apathy may reside in our innate cognitive biases. One of these biases is "hyperbolic discounting." Humans are not known for thinking about the long term. Evolutionarily, short-term attention to immediate threats was more valuable than long-term perspectives. During the early days of evolution, humans were mostly challenged with short-term threats such as predators and natural disasters. The lack of concern for future generations, and the refusal to sacrifice current comforts in order to avoid future harm, may be a derivative of this bias. Another cognitive bias responsible for our passive attitude is the bystander effect. We tend to believe that someone else (usually our leaders) will do the work for us. The greater the number of bystanders, the less likely one of them will assist a person in distress.

These cognitive biases are the consequence of deeper, more profound reasons for the lack of a collective will to address this crucial problem. What are those reasons? How can we motivate people to act?

The first answer that comes to my mind, and one that represents the complexity and depth of the problem, is anthropocentrism, assigning intrinsic value to human beings alone.[163] We do not know the exact turning point that started our gradual departure from nature. It likely occurred with the emergence of symbolic communication tens of thousands of years ago.[164] To grasp the change that humans have undergone over time, consider the difference between the representations of human figures in the upper Paleolithic period (between 50,000 and 12,000 years ago) to those of today. In ancient cave paintings from Lascaux, France, human figures have no face. Today, we live in the age of selfies. In other words, we have moved from the solidarity of small groups of hunter-gatherers who lived sustainably to the spread of narcissism in modernity, where the individual human is center stage.

> This rock painting, dating to 6,000–2,000 years ago (late Stone Age period), shows a giraffe, antelopes, and a spotted lion with human toes. The hybrid lion-man figure was probably used for shamanistic rituals. Twyfelfontein, Namibia.

163 In ethics, the intrinsic value of something is said to be the value that thing has on its own.

164 The marker of the exact turning point is under scientific debate. Many scientists believe that a significant leap in the cultural skills of *Homo sapiens* took place around 70,000 years ago, with the development of symbolic language. Probably unlike animals, humans can imagine and speak about things that do not really exist. This characteristic is one of the foundations of human society and of our ability to cooperate and collectively address the constraints of natural selection.

Anthropocentrism is the foundation of Western culture, and deeply rooted in Anthropocentrism is the foundation of Western culture, and deeply rooted in our psyche. The chasm between humans and nature, which has expanded over time, is the root of our inability to feel deep empathy towards all living creatures and nature. Moreover, the process of continuous disenchantment from nature increasingly facilitates our domination over it.

Ever since humans became an abundant species, biodiversity has been deeply altered on many levels, including the mixing of species and their habitats. An example of this is the Nile Valley, which was cultivated by ancient Egyptians a few thousand years ago and has never returned to its primordial swamp landscape. In a massive global assessment of land use conducted by the ArchaeoGLOBE project,[165] it was found that by 3,000 years ago, Earth's terrestrial ecology had been significantly affected by hunter-gatherers, pastoral farming and growing crops. These results echo the findings presented by William M. Denevan in his 1992 article, "The Pristine Myth: The Landscape of the Americas in 1492." In it, Denevan argues that before Columbus's 1492 arrival in America, the area was not actually "a world of barely perceptible human disturbance" but was actually a "humanized landscape almost everywhere."

Still, something was significantly different in ancient times. Animism was prevalent and people attributed spiritual essence to places, objects and animals. Every tree, every spring, every river had its guardian deity. People could not exploit nature without limit, and without placating the local deity. To a certain extent, this meant that protection of the environment was self-regulated and embedded in people's beliefs, something which enabled sustainable living.

Anthropocentrism led to significant changes in human thought, and these have had an immense impact on our relationship with nature. A giant leap towards the separation between humans and nature took place two thousand years ago, with the birth of Judeo-Christian theology. As American historian Lynn White (1907–1987) asserts in his influential 1967 lecture, Judeo-Christian theology, which replaced pagan animism in the West, established the dualism of man and nature, and the concept that humans are superior to nature and entitled to exploit it for our benefit. It is sufficient to read the story of creation in Genesis 1:26 and 1:28 to understand how deeply this theology dictates our relationships with nature:

> *And God said, Let us make man in our image, after our likeness: and let them have dominion over the fish of the sea, and over the*

165 https://archaeoglobe.com/ (retrieved March 13, 2021).

> *fowl of the air, and over the cattle, and over all the earth, and over every creeping thing that creepeth upon the earth. And God blessed them, and God said unto them, Be fruitful, and multiply, and replenish the earth, and subdue it: and have dominion over the fish of the sea, and over the fowl of the air, and over every living thing that move the upon the earth.*

White also claims that science was historically aristocratic and intellectual, while technology was lower class and empirical. The democratic revolution decreased the social barrier of entry into science, fusing science and technology and perpetuating humans' impact on the environment. The severe consequences of technology over the past four generations can be seen everywhere and have reached a scale that puts the global atmosphere at risk.

The decrease in the centrality of religion over the past century has not diminished anthropocentrism. Social values today are still based on the pillars of Judeo-Christian theology. Furthermore, we have not truly embraced the central message of evolution theory: humans are part of nature.

From all of the above, it is clear to me that the future of our planet and humanity is profoundly dependent upon our perspective of the human–nature relationship. Science and technology are more of an enabler, and less of a solution, to the present ecological crisis.

So how can we overcome our passivity? How can we drive humanity to action? Historically, the first attempt was to promote shallow ecology, which means fighting pollution and resource exploitation to benefit the health and affluence of people, mainly in developed countries. This attempt has profoundly failed to resolve the abuse of natural resources. In many cases, massive "nature conservation" campaigns have been ineffective, and most people remain indifferent to the continuous destruction of natural habitats.

I firmly believe that we need a paradigm change, perhaps in the form of a new eco-religion that will bind people together for a cause, and call humanity to action. Seeds for such a change have accompanied human thought for many years, and there are positive early signs that modern thinking is moving towards a more holistic view of the relationships between humans and nature. Scientific and metaphysical ideas such as complex systems theory, some interpretation of quantum mechanics, various new studies that highlight the possibility of consciousness in non-human creatures, shifts in the focus of evolution theory towards the influence of the environment on phenotypes, the adoption of Engaged Buddhism,[166] the Gaia hypothesis, and many more. Although these ideas may seem unrelated on the face of it, they are a part of the spirit of the times.

Such a paradigm change, which is still considered radical, is deep ecology, developed during the 1970s by Scandinavian philosophers, notably Arne Næss (1912–2009). Deep ecology broadly sees the environmental crisis in the context of sexism, racism, heterosexism, neoliberalism, and other forms of hidden or explicit dominations over others. As opposed to anthropocentrism, Næss and his colleagues proposed a relational total-field image of the world. They argue that the basic concept of environmental ethics is that humans and Earth's biota each have their own intrinsic value. Based on this ecosophical perspective, humans, like other organisms, are just one strand in a greater biotic interacting web. Therefore, conceptualizing the links between humanity and other organisms will result in greater empathy towards nature and the world.

Deep ecology is an appealing way of thinking and might be part of human thought's evolutionary course. However, it contradicts humans' deeply rooted anthropocentric view, and therefore its impact might take more time than we have. We need to think about how to accelerate the adoption of this perspective so peoples' state of mind can be changed before it is too late.

One potential direction is investing more effort in increasing peoples' awareness of animal consciousness. If we extend our "borders of consciousness," we might create a turning point in how we perceive human–nature relationships.

French existential philosopher Emmanuel Levinas (1906–1995) called it the capacity to see "the face of the other," and this may be a prerequisite for including animals in our moral community. Dogs, and to some extent other popular, aesthetic animals have it, and it directly influences how people morally view them and their welfare.

For all practical purposes, sentient animals in general, and sentient marine animals in particular, should be included within humanity's moral demarcation line. We should stop treating animals as inferiors. Many animals have the capacity for pleasure and pain, something we tend to ignore through our culturally trained cognitive dissonance. By denying that animals are sentient, people reduce their moral discomfort; if animals don't suffer, we can continue treating them cruelly.

166 Engaged Buddhism is a movement that entails Buddhist values, perspectives, and insights to cope with social, political and environmental crises. One of the thought leaders of this approach is Vietnamese Buddhist teacher Thích Nhât Hanh.

Animals do have consciousness, and some even have developed consciousness. This is not a new idea; it was part of Aristotle's theory of nature 2,400 years ago,[167] and discussed by Darwin 160 years ago. In the words of animal rights activist and brilliant philosopher Peter Singer (1946–), "All the arguments to prove man's superiority cannot shatter this hard fact: in suffering, the animals are our equals."[168]

I do not suggest promoting the "humanization" of sea creatures; on the contrary, I suggest further investing in research to establish the foundation to educate people to accept other possible minds as part of the natural way in which we view the world. This will enable people to adopt a more tolerant approach, which means that alongside the uniqueness of each and every human being, we are not independent and are, in fact, tightly related to other creatures.

In other words, identification between the human ego and nature will expand the boundaries of the self beyond our physical body.[169] This might serve as an effective way of mobilizing people and making them emotionally engaged with a large number of animals. It is challenging; however, it is also in line with today's evolving attitude towards nature and animals, as reflected in the growing mainstream popularity of vegetarianism and veganism.

I would like to end on an optimistic note. It is essential to understand that we still have a good chance of preserving these magnificent marine habitats. Compared with terrestrial habitats, the marine environment was relatively immune to human impact until recent centuries. The defaunation process of marine habitats began thousands of years after it began in terrestrial habitats. Thus, marine fauna and flora are in a relatively better condition than terrestrial equivalents. Humans are responsible for many ecological and local habitat extinctions. Nevertheless, and unlike terrestrial animal extinctions, human interference is responsible for only a few complete marine animal extinctions. Therefore, due to the high mobility of many marine animals, rehabilitation of aquatic habitats is more plausible and within reach. So, acting to slow defaunation might yield a positive result that might be felt even in the short term, and which might encourage people and call humanity to action.

The Buddhist metaphor of Indra's infinite jeweled net wonderfully represents a mutual identity worldview, one which contrasts with the Judeo-Christian tradition. Like the jewel at every node that reflects all the other jewels in this cosmic matrix, all

---

167 Aristotle categorizes the soul into three hierarchical categories: the first layer is the plant's nutritive and reproductive soul; the second layer is demonstrated by the elephant's sensitive, motor and sensory soul; and the third layer is the human's rational soul.

168 From a profile about him published in *The Guardian* in February, 2009.

169 https://plato.stanford.edu/entries/ethics-environmental/#IntChaEnvEth (retrieved January 17, 2021).

things in the universe are interconnected and interrelated. Every jewel metaphorically represents an atom, cell, organism or unit of consciousness that is connected to all other things. Coral reef ecosystems are an excellent example of such interconnected systems. The moral embedded in this perspective of Indra's net is that compassion must be extended to all living things. All human actions cause a ripple effect and should, therefore, be ethically considered subject to this worldview.

˅ My son, Or, diving behind black coral. Sinai Peninsula, Egypt.

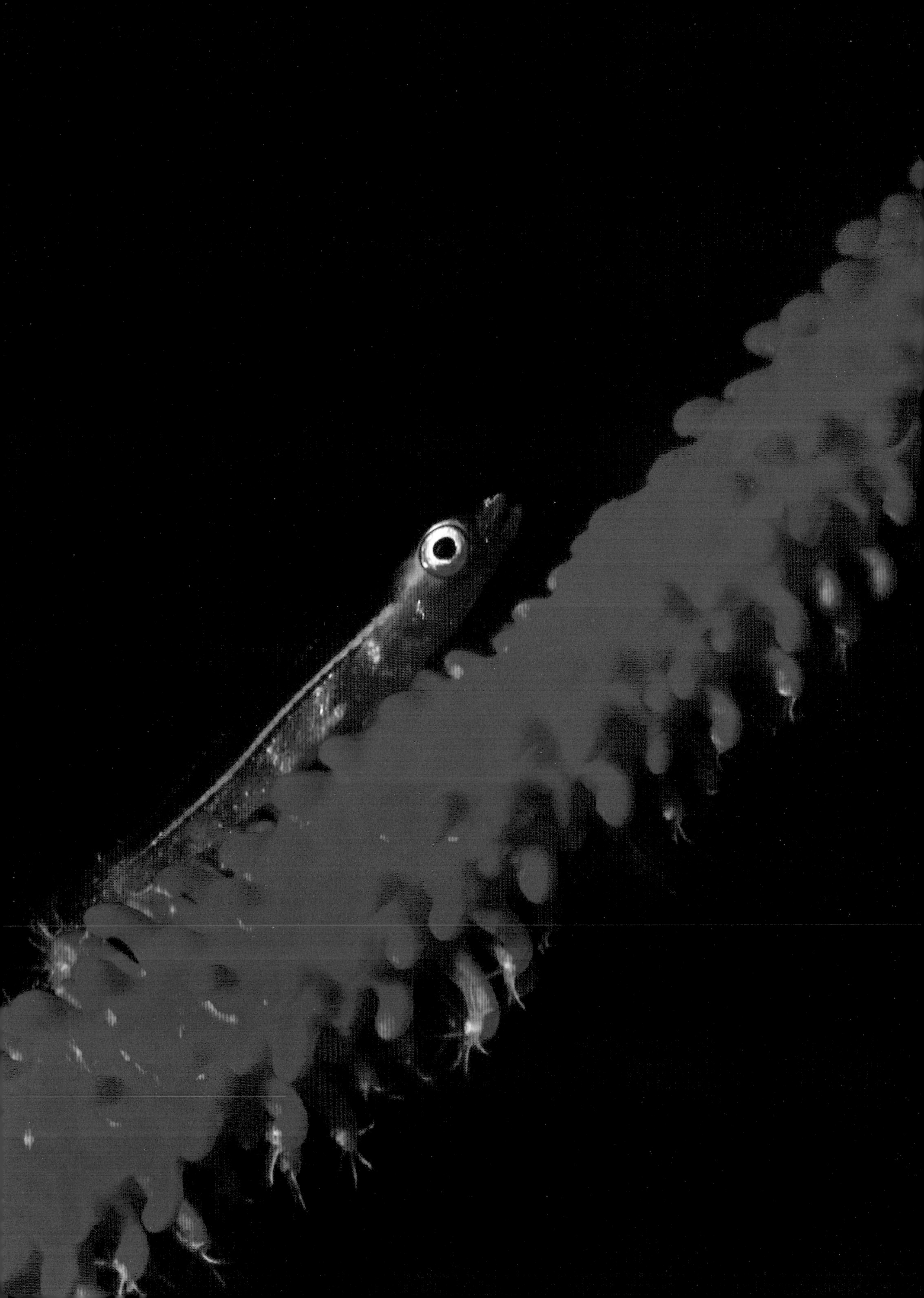

Afterword

By Dr. Tom Shlesinger

Marine Ecologist, Institute for Global Ecology,
Florida Institute of Technology, Florida, USA

Reflecting on Reflections

« A whip coral goby (*Bryaninops yongei*). Puerto Galera, Philippines.

Many books about nature try to be "objective," taking the human experience out of the equation and turning it into pure information. By contrast, the essence of *Reflections Underwater* is just the opposite – in many ways, the book places human experience, perception and history in the center of the debate about how and why we interpret particular things in particular ways.

I first became aware of *Reflections Underwater* on a beautiful sunny day in Florida in May 2020. I was isolated in my house, staring at the vast Atlantic Ocean, when I received an email from Oded Degany. This highly respected corporate executive was seeking a consultant for a book he was writing about (no less than) tropical coral reefs. His message immediately intrigued me as he described the cross-disciplinary course he wished to take with his book; a journey through many topics relating to evolution, ecology, behavior, mathematics, physics, psychology, philosophy, arts and how they are all manifested in the nature of coral reefs. I remember thinking – what makes a person who is an expert in strategic planning, corporate merger and acquisition, want to devote so much time to writing such a book? Is it the desire of an avid diver, who witnessed and photographed many of the most incredible underwater sights, to publish a photography book? Or is it a consequence of the infamous 2020 that caused us all be isolated at home under lockdown, looking for something to do? Will this be another kind of zoological textbook, summarizing what is already known? Will it try to touch, perhaps, too many of the complex topics pertaining to the science of the oceans and risk becoming a superficial compilation of facts and anecdotes?

Shortly thereafter, we had our first video-chat meeting that evolved into a lively discussion which went on for hours. We started talking about coral reefs and quickly arrived at anthropological and philosophical aspects of the human experience and perception, while always circling back to coral reefs and the many mysteries they harbor. Oded's diverse personal, professional, and educational background lead to a fascinating fusion of terms and views from completely different disciplines and schools of thought. In this book, Oded borrows perspectives and theories from physics and mathematics to ponder the biology and ecology of coral reefs. This approach, alongside his natural curiosity about human culture and the nature that lies beneath the ocean surface, all make Oded's point of view quite unique and in some cases, even a bit controversial. And that is exactly what is special about this book. It reminds the reader that science has yet to have its final say on many topics because so much is still undiscovered or not fully understood.

While this book tells some of the epics of the underwater world with many of the animals and phenomena beautifully illustrated by Oded's photos, this is not its only virtue. If you, the reader, have reached this point and are reading these lines, you have probably already figured it out yourself. By discussing both historic and contemporary, scientific and philosophic understandings, but also some of their alternatives, this book not only provides you with well-told and illustrated facts about nature but, in fact, strives to challenge and to provoke you to think. Whether you are an expert in some field of marine science, a well-traveled and experienced diver who knows a lot about this environment, or a "newbie" interested in the underwater world, I am sure this book introduced you to some new and fascinating ideas and phenomena.

Though I cannot know where this book leaves you, I can tell you where it leaves me. As a marine ecologist who has devoted the last two or so decades of his life exploring the magnificent world created by corals, I find this book very intriguing, full of interesting nature and science stories that made me go further and search for more details about them. In fact, even as somewhat of a coral-reef "expert," this book sparked in my mind several new research avenues, which I had not thought about before and which I intend to follow in the years to come.

Lastly, I would like to touch upon the burning environmental crisis we are facing. That humans are causing so many changes to our planet is well known. The current geological epoch is commonly referred to as "The Anthropocene" because of this understanding. That many populations and whole ecosystems, especially in the ocean, are on the brink of a catastrophic collapse is also commonly discussed. Yet, we are still failing to address some of the greatest contemporary threats, such as the biodiversity and climate crises, which have a lot to do with mankind's excess levels of consumption, production, pollution, etc. Many more people around the world must become more aware of the direction our planet is headed before we drive it to places where we do not want to be. It is my belief that works like this book, among many other initiatives, are imperative to that goal, as knowledge eventually evokes understanding, understanding promotes compassion, and compassion leads to care and action.

March 2021

Concise Glossary

Throughout the book, most of the professional terms are defined within the text. Below are the most useful terms, in simple language, for readers who are not familiar with relevant scientific terminology. Although most terms appear in the footnotes, I recommend becoming familiar with these shortlisted terms, as they frequently appear in the book.

Algae (Zooxanthellae) Collective term for single-cell (eukaryote) photosynthetic organisms that are part of the phylum Dinoflagellate. Algae live in symbiosis with diverse coral reef animals such as sea anemones, nudibranchs and bivalves. The most prominent symbiotic example is with corals, where algae produce more than 90 percent of their required food.

Anthropocentrism A philosophical perspective that considers human beings to be the world's central entity, separate and superior to nature.

Aposematism Animal signaling to potential prey that it is poisonous or unprofitable. A primary form of aposematism is conspicuous, high-contrast patterns of bright colors (warning coloration).

Asexual reproduction A form of reproduction in which an organism can reproduce in the absence of a mating partner. Asexual reproduction involves a single parent and there is no exchange of genes with another parent; the result is offspring that are genetically identical to one another and to the parent. Budding is a form of asexual reproduction.

Batesian mimicry A type of mimicry in which a harmless creature has evolved to mimic a harmful creature's warning signals without any accompanying defense.

Bioluminescence Light produced in the body of organisms as a result of a chemical reaction between luciferin, a light-emitting molecule, and luciferase, an enzyme which differs between species. Bioluminescence can be produced by the organisms or as part of symbiosis with bacteria.

Budding An asexual form of reproduction in which an organism develops from a bud that later becomes an independent, genetically identical organism. Coral colonies are formed and grow by budding.

Cambrian explosion An event that occurred around 542 million years ago marked by the Big Bang of biological diversity and a significant change in the variety, diversification and complexity of life forms. Many of today's animals originated and diversified during the Cambrian evolutionary radiation.

Cephalopods Members of the class Cephalopoda of the phylum Mollusca. A small group of highly mobile and intelligent animals that live exclusively in marine habitats. The most familiar representatives of this class are octopuses, squids, cuttlefishes and chambered nautiluses. The most significant feature of cephalopods is their possession of eight to ten tentacles or arms (up to 90 in the nautilus).

Chromatophores Pigment-containing, light-reflecting cells present in various animals, in a deep layer of their skin.

Cnidaria Phylum characterized by specialized stinging cells called cnidocytes (or nematocytes), usually with radial symmetry. Common animals in this phylum are corals, sea anemones, sea pens, jellyfishes and the Portuguese man o' war.

Complex system A system composed of many interacting components whose behavior is difficult to predict due to the interdependencies between its parts and its environment. Its properties are not fully explained by

« Circular spadefish (*Platax orbicularis*). Puerto Galera, Philippines.

its parts, but by distinct emergent properties that arise from the interaction between its parts. Examples of such properties include nonlinearity, self-organization, adaptation and feedback loops.

Convergent evolution Unrelated organisms that independently evolve similar (analogous) structures, traits and features such as body form, coloration and organs. For example, similarities between the hydrodynamic structures of fishes and dolphins, although these two creatures do not share a common ancestor.

Coral colony Group of connected, partially integrated, genetically identical polyps that live in close association. A coral colony is a result of asexual reproduction (budding).

Crustaceans A large group of invertebrates belonging to phylum Arthropoda; includes mantis shrimps, crabs and lobsters.

Crypsis An animal's ability to avoid detection by other animals by disguising its appearance.

Divergent evolution A process in which two different species share a common ancestor but have evolved and accumulated different characteristics. Similarities between species explained by divergence from a common ancestor are called homologous structures.

Eukaryote Organisms whose cells have a nucleus enclosed within a membrane (eukaryotic cells).

Fluorescence A process of re-emission of light by a substance immediately following the incident light's absorption. In most cases, the emitted light color is different from the incident light.

Genotype An individual's heritable collection of genes.

Hermaphroditism The expression of both female and male sex organs during an individual's lifetime.

Holobiont A composition of different species that live in symbiosis and create single ecological units. Humans are considered holobionts since we have bacteria, viruses, and sometimes carry larger organisms such as worms and fungi.

Hybridization The production of offspring through the interbreeding of individuals of different species.

Larval stage A developmental stage in the life cycle of some animals that juveniles undergo before metamorphosing into adults. This stage is characterized by different structures and organs that change while transforming into an adult.

Mimicry An evolved physical or behavioral resemblance between one organism to another, usually providing a selective advantage to the mimicker.

Mollusca A phylum of soft-bodied invertebrates, many of which secrete protective dorsal shells of calcium carbonate. Examples include octopuses, sea slugs, nudibranchs and clams.

Müllerian mimicry A form of biological resemblance in which two or more unrelated, unpalatable organisms exhibit similar warning systems. It reduces the mortality involved in training predators to avoid these organisms because fewer experiences of encountering members of either relevant species are required for learning. Müllerian mimicry differs from Batesian mimicry because all species involved are distasteful to predators.

Phenotype An organism's observable physical characteristics, such as morphology, developmental process and behavior. The phenotype is mainly derived from the genotype expression mechanism and the influence of the environment.

Photoreceptor A special sensory cell in the eye's retina that converts incoming light properties into electric signals that are carried to the brain, where they are interpreted.

Photosynthesis The process of converting carbon dioxide, water and light energy, into sugars and oxygen ($6CO_2$ + $6H_2O$ + photons (sunlight) $\rightarrow C_6H_{12}O_6 + 6O_2$). It is one of life's most extraordinary and fundamental processes.

Phylum A biological taxon that comes third in the hierarchy of classification, between kingdom and class. There are 35 animal phyla, characterized by the same distinctive body plan for each of the classes that form it.

Plankton A diverse collection of organisms with limited swimming capabilities that drift with the current and live in large bodies of water. There are three functional groups of plankton: phytoplankton are plant-like organisms that manufacture their own food using energy from sunlight; bacterioplankton are bacteria; zooplankton are tiny animals such as larvae.

Planula larva A very young, free-swimming, larval form, common in many species of cnidarians.

Polarized light waves Light waves in which the electric field's vibration is well-defined and restricted to a single plane. Light is unpolarized if the direction of this electric field vibrates randomly. Lasers are an example of polarized light, while the sun and LED spotlights are examples of unpolarized light.

Polymorphism In the context of cnidarians, polymorphism could be the existence of individuals (zooids) of a single species, in more than one physiological and morphological form and in different functions, that act in a well-organized division of labor. For example, the Portuguese man o' war is a colony of specialized, genetically identical zooids with diverse forms and functions, all working together as a single organism.

Polyp A tiny, soft-bodied organism that functions as the coral's basic unit. Polyps can live individually, as they do in mushroom corals, or in large colonies, as they do in reef-building corals.

Positive feedback loop A process in which the end products of an action reinforce an increase in that action (when A produces more of B, which produces more of A and so forth). In contrast to negative feedback loops, positive feedback loops intensify changes and tend to move a system away from its initial state. Positive feedback has driven many of the most significant changes in evolution and ecology. Feedback loops (both negative and positive ones) allow living organisms to maintain homeostasis, i.e., keep the internal environment steady.

Sexual reproduction A process involving the fusion of reproductive cells. The offspring is genetically different from the parents. Sexual reproduction increases genetic variation and facilitates adaptation to environmental challenges.

Speciation An evolutionary process by which populations diverge and form distinct new species.

Superorganism A tightly integrated collection of single creatures from the same species that work as a single, coordinated, self-sustaining biological entity. The colony comprises individuals specializing in a single function (labor division) that collectively perform like an organism.

Symbiosis An evolved cooperation or close living relationship between organisms from distinct species, with benefits to at least one of the organisms involved.

Teleosts The largest group of fishes, including more than 95 percent of all present-day fishes.

Tinkering In the context of evolution, tinkering is the creation of evolutionary novelty through random combinations of available, preexisting resources. A tinkerer works without a clear plan in mind, collecting everything at its disposal, and rearranging these components to produce a workable object.

Umwelt An organism's self-centered, subjective perceptual world.

⌄ A fire urchin. Puerto Galera, Philippines.

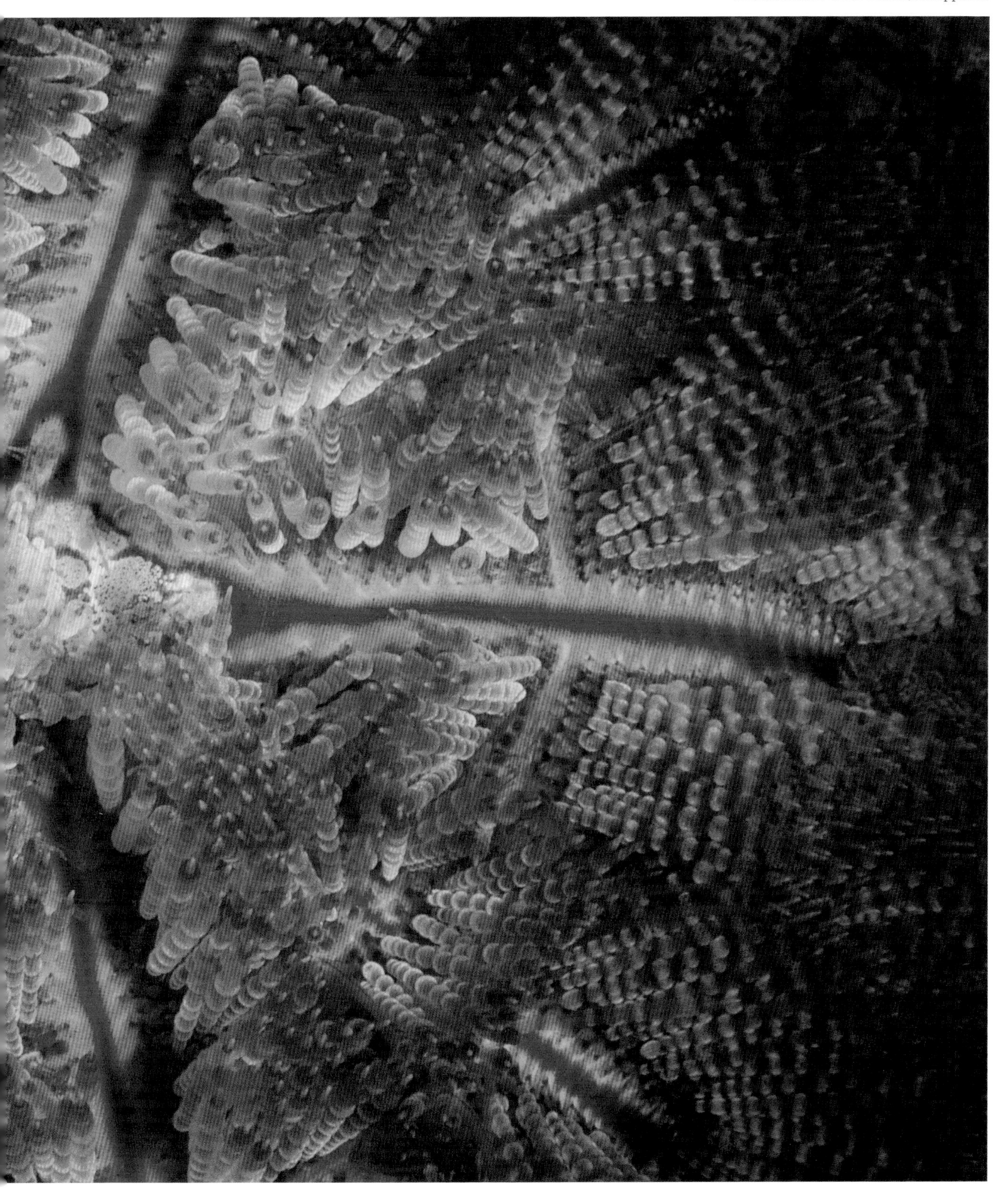

Bibliography

General literature

Dubinsky, Z., & Stambler, N. (eds.) (2010). *Coral Reefs: An Ecosystem in Transition.* Springer Science & Business Media.

Humann, P., & DeLoach, N. (2010). *Reef Creatures Identification: Tropical Pacific* (2nd edn.). New World Publications, Inc.

Humann, P., & DeLoach, N. (2010). *Reef Creature Identification: Tropical Pacific.* New World Publications.

Opening

Verne, J. (1998). *Twenty Thousand Leagues under the Sea.* Oxford University Press. (Original work published 1869.)

Prologue

Riefenstahl, L. (2003). *Impressionen unter Wasser.* EuroVideo.

Chapter 1

The Uniqueness of Diving Experience

Pirsig, R. M. (1999). *Zen and the Art of Motorcycle Maintenance: An Inquiry into Values.* Random House. (Original work published 1974.)

Sensual Experience and the Enigma Behind Our Attraction to Diving

Beneton, F., Michoud, G., Coulange, M., Laine, N., Ramdani, C., Borgnetta, M., Breton, P., Guieu, R., Rostain, J. C., & Trousselard, M. (2017). Recreational diving practice for stress management: An exploratory trial. *Frontiers in Psychology*, 8(2193).

Clayton, L. W. (2003). *Identity and the Natural Environment: The Psychological Significance of Nature.* MIT Press.

Cooper, J. C. (2012). *An Illustrated Encyclopaedia of Traditional Symbols.* Thames & Hudson.

Dutton, D. (2014). A Darwinian theory of beauty. *Philosophy and Literature*, 38(1), A314–A318.

Eliade, M. (1959). *The Sacred and the Profane: The Nature of Religion.* Houghton Mifflin Harcourt.

Foley, R., & Lahr, M. M. (2014). The role of "the aquatic" in human evolution: Constraining the aquatic ape hypothesis. *Evolutionary Anthropology: Issues, News, and Reviews*, 23(2), 56–59.

Gislén, A., Dacke, M., Kröger, R. H., Abrahamsson, M., Nilsson, D. E., & Warrant, E. J. (2003). Superior underwater vision in a human population of sea gypsies. *Current Biology*, 13(10), 833–836.

Ilardo, M. A., Moltke, I., Korneliussen, T. S., Cheng, J., Stern, A. J., Racimo, F. et al. (2018). Physiological and genetic adaptations to diving in sea nomads. *Cell*, 173(3), 569–580.

Jablonka, E., & Lamb, M. J. (2014). *Evolution in Four Dimensions: Genetic, Epigenetic, Behavioral, and Symbolic Variation in the History of Life* (revised edition). MIT Press.

<< A feather star.
Bunaken, Indonesia.

Jung, C. G. (1990). *The Archetypes and the Collective Unconscious* (trans. R. F. C. Hull). Princeton University Press. (Original work published 1959.)
Kahn Jr., P. H. (1997). Developmental psychology and the biophilia hypothesis: Children's affiliation with nature. *Developmental Review*, 17(1), 1–61.
Kellert, S. R. (1993). The biological basis for human values of nature. In S. R. Kellet & E. O. Wilson (eds.), *The Biophilia Hypothesis*. Island Press, 42–69.
Lindholm, P., & Lundgren, C. E. (2009). The physiology and pathophysiology of human breath-hold diving. *Journal of Applied Physiology*, 106(1), 284–292.
Lumsden, C. J., & Wilson, E. O. (2005). *Genes, mind, and culture – The coevolutionary process*. World Scientific.
Melville, H. (1988), *Moby-Dick; or, the Whale, The Writings of Herman Melville* (Vol. 6) (Harrison Hayford, Hershel Parker, and G. Thomas Tanselle, eds.) Northwestern University Press and the Newberry Library. (Original work published 1851.)
Morgan, E. (1982). *The aquatic ape*. Scarborough House.
Nichols, W. J. (2014). *Blue mind: The surprising science that shows how being near, in, on, or under water can make you happier, healthier, more connected, and better at what you do*. Little, Brown and company.
Rae, T. C., & Koppe, T. (2014). Sinuses and flotation: Does the aquatic ape theory hold water? *Evolutionary Anthropology: Issues, News, and Reviews*, 23(2), 60–64.
Schagatay, E. (2011). Human breath-hold diving ability suggests a selective pressure for diving during human evolution. In M. Vaneechoutte, A. Kuliukas & M. Verhaegen (eds.), *Was Man More Aquatic in the Past? Fifty Years after Alister Hardy Waterside Hypotheses of Human Evolution*. Bentham Books, 120–147.
Shiota, M. N., Keltner, D., & Mossman, A. (2007). The nature of awe: Elicitors, appraisals, and effects on self-concept. *Cognition and emotion*, 21(5), 944–963.
Shoham, S. G. (2004). *The myth of Tantalus: A scaffolding for an ontological personality theory*. Sussex Academic Press.
Time Staff (2003, September 12), Jacques Cousteau, *Time*, http://content.time.com/time/specials/packages/article/0,28804,1981290_1981333_1981618,00.html (retrieved March 5, 2021).
Wilson, E. O. (1984). *Biophilia*. Harvard University Press.

Intellectual Experience

Barthes, R. (1981). *Camera lucida: Reflections on photography*. Macmillan.
Baudrillard, J. (1994). *Simulacra and simulation*. University of Michigan Press.
Darimont, C. T., Codding, B. F., & Hawkes, K. (2017). Why men trophy hunt. *Biology Letters*, 13(3), 20160909.
Kahneman, D. (2011). *Thinking, fast and slow*. Macmillan.
Sontag, S. (1990). *On Photography*. Anchor-Doubleday. (Original work published 1977.)
Whitman, W. (1865). *Leaves of grass*. D. McKay.
Žižek, S. (2013). *Welcome to the desert of the real: Five essays on September 11 and related dates*. Verso Trade.

Chapter 2
Coral Reefs – A Holistic View

Darwin, C. (1842). *The Structure and Distribution of Coral Reefs: Being the First Part of the Geology of the Voyage of the Beagle, under the command of Capt. Fitzroy, RN During the Years 1832–1836*. Smith, Elder.

Darwin, C. (1859). *On the Origin of Species by Means of Natural Selection Or the Preservation of Favoured Races in the Struggle for Life.* H. Milford; Oxford University Press.
Fisher, R., O'Leary, R. A., Low-Choy, S., Mengersen, K., Knowlton, N., Brainard, R. E., & Caley, M. J. (2015). Species richness on coral reefs and the pursuit of convergent global estimates. *Current Biology,* 25(4), 500–505.
Garwood, R. J., & Edgecombe, G. D. (2011). Early terrestrial animals, evolution, and uncertainty. *Evolution: Education and Outreach,* 4(3), 489–501.
Levin, S. A. (2001). Editor-in-Chief. *Encyclopedia of Biodiversity* (p. 1). Elsevier.
Limmon, G., Delrieu-Trottin, E., Patikawa, J., Rijoly, F., Dahruddin, H., Busson, F. et al. (2020). Assessing species diversity of Coral Triangle artisanal fisheries: A DNA barcode reference library for the shore fishes retailed at Ambon harbor (Indonesia). *Ecology andEevolution,* 10(7), 3356–3366.
Primack, R. B. (2002). *Essentials of conservation biology.* Sinauer Assoc. Inc.
Reaka-Kudla, M. L. (1997). The global biodiversity of coral reefs: a comparison with rain forests. In M. Reaka-Kudla, D. E. Wilson & E. O. Wilson (eds.), *Biodiversity II: Understanding and protecting our biological resources.* Joseph Henry Press, 83–108.
Spalding, M. D., Ravilious, C., & Green, E. P. (2001). *World atlas of coral reefs.* University of California Press.
Wallace, A. R., (1890). *The Malay Archipelago.* Macmillan and Co.

What Are Corals?

Bowen, J. (2015). *The Coral Reef Era: From Discovery to Decline: A history of scientific investigation from 1600 to the Anthropocene Epoch.* Springer.
Cowman, P. F. (2014). Historical factors that have shaped the evolution of tropical reef fishes: a review of phylogenies, biogeography, and remaining questions. *Frontiers in genetics,* 5, 394.
Dubinsky, Z., & Stambler, N. (eds.). (2010). *Coral reefs: an ecosystem in transition.* Springer.
Egerton, F. N. (2012). *Roots of ecology: antiquity to Haeckel.* University of California Press.
Groombridge, B., Jenkins, M. D., & Jenkins, M. (2002). *World atlas of biodiversity: earth's living resources in the 21st century.* University of California Press.
Harrison, P. L. (2011). Sexual reproduction of scleractinian corals. In Z. Dubinsky & N. Stambler (eds.), *Coral reefs: an ecosystem in transition.* Springer, 39–85.
Kleypas, J. A., McManus, J. W., & Menez, L. A. (1999). Environmental limits to coral reef development: where do we draw the line? *American zoologist,* 39(1), 146–159.
LaJeunesse, T. C., Parkinson, J. E., Gabrielson, P. W., Jeong, H. J., Reimer, J. D., Voolstra, C. R., & Santos, S. R. (2018). Systematic revision of Symbiodiniaceae highlights the antiquity and diversity of coral endosymbionts. *Current Biology,* 28(16), 2570–2580.
Levitan, D. R., Fukami, H., Jara, J., Kline, D., McGovern, T. M., McGhee, K. E. et al. (2004). Mechanisms of reproductive isolation among sympatric broadcast-spawning corals of the Montastraea annularis species complex. *Evolution,* 58(2), 308–323.
Loya, Y., & Sakai, K. (2008). Bidirectional sex change in mushroom stony corals. *Proceedings of the Royal Society B: Biological Sciences,* 275(1649), 2335–2343.
Shlesinger, T., & Loya, Y. (2019). Breakdown in spawning synchrony: A silent threat to coral persistence. *Science,* 365(6457), 1002–1007.
Shlesinger, T., & Loya, Y. (2019). Sexual reproduction of scleractinian corals in mesophotic coral ecosystems vs. shallow reefs. In Y. Loya, K. A. Puglise & T. C. L. Bridge (eds.), *Mesophotic Coral Ecosystems.* Springer, 653–666.
Van Oppen, M. J., Souter, P., Howells, E. J., Heyward, A., & Berkelmans, R. (2011). Novel genetic diversity through somatic mutations: fuel for adaptation of reef corals? *Diversity,* 3(3), 405–423.
Wild, C., Huettel, M., Klueter, A., Kremb, S. G., Rasheed, M. Y., & Jørgensen, B. B. (2004). Coral mucus functions as an energy carrier and particle trap in the reef ecosystem. *Nature,* 428(6978), 66–70.

Biodiversity in the Coral Triangle

Barber, P. H. (2009). The challenge of understanding the Coral Triangle biodiversity hotspot. *Journal of Biogeography*, 36(10), 1845–1846.

Bellwood, D. R., Renema, W., & Rosen, B. R. (2012). Biodiversity hotspots, evolution and coral reef biogeography. In D. Gower, K. Johnson, J. Richardson, B. Rosen, L. Rüber & S. Williams (eds.), *Biotic evolution and environmental change in Southeast Asia*, Cambridge University Press, 216–245.

Brown, J. H. (2014). Why are there so many species in the tropics? *Journal of biogeography*, 41(1), 8–22.

Burke, L., Reytar, K., Spalding, M., & Perry, A. (2011). *Reefs at risk revisited*. World Resources Institute.

Hardin, G. (1960). The competitive exclusion principle. *Science*, 131(3409), 1292–1297.

Huang, D., Goldberg, E. E., Chou, L. M., & Roy, K. (2018). The origin and evolution of coral species richness in a marine biodiversity hotspot. *Evolution*, 72(2), 288–302.

Kiessling, W., Simpson, C., & Foote, M. (2010). Reefs as cradles of evolution and sources of biodiversity in the Phanerozoic. *Science*, 327(5962), 196–198.

Miller, E. C., Hayashi, K. T., Song, D., & Wiens, J. J. (2018). Explaining the ocean's richest biodiversity hotspot and global patterns of fish diversity. *Proceedings of the Royal Society* B, 285(1888), 20181314.

Richmond, R. H. (1987). Energetics, competency, and long-distance dispersal of planula larvae of the coral *Pocillopora damicornis*. *Marine Biology*, 93(4), 527–533.

Rohde, K. (1992). Latitudinal gradients in species diversity: the search for the primary cause. *Oikos*, 514–527.

Thornhill, D. J., Mahon, A. R., Norenburg, J. L., & Halanych, K. M. (2008). Open-ocean barriers to dispersal: a test case with the Antarctic Polar Front and the ribbon worm *Parborlasia corrugatus* (Nemertea: Lineidae). *Molecular ecology*, 17(23), 5104–5117.

Veron, J. E., Devantier, L. M., Turak, E., Green, A. L., Kininmonth, S., Stafford-Smith, M., & Peterson, N. (2009). Delineating the coral triangle. *Galaxea, Journal of Coral Reef Studies*, 11(2), 91–100.

Veron, J., Stafford-Smith, M., DeVantier, L., & Turak, E. (2015). Overview of distribution patterns of zooxanthellate Scleractinia. *Frontiers in Marine Science*, 1, 81.

What Does a Pristine Coral Reef Look Like?

Bradley, D., Conklin, E., Papastamatiou, Y. P., McCauley, D. J., Pollock, K., Pollock, A. et al. (2017). Resetting predator baselines in coral reef ecosystems. *Scientific Reports*, 7, 43131.

Díaz-Pérez, L., Rodríguez-Zaragoza, F. A., Ortiz, M., Cupul-Magaña, A. L., Carriquiry, J. D., Ríos-Jara, E. et al. (2016). Coral reef health indices versus the biological, ecological and functional diversity of fish and coral assemblages in the Caribbean Sea. *PLoS One*, 11(8), e0161812.

Faurby, S., Silvestro, D., Werdelin, L., & Antonelli, A. (2020). Brain expansion in early hominins predicts carnivore extinctions in East Africa. *Ecology Letters*.

Friedlander, A. M., Sandin, S. A., DeMartini, E. E., & Sala, E. (2010). Spatial patterns of the structure of reef fish assemblages at a pristine atoll in the central Pacific. *Marine Ecology Progress Series*, 410, 219–231.

Gordon, T. A., Radford, A. N., Davidson, I. K., Barnes, K., McCloskey, K., Nedelec, S. L. et al. (2019). Acoustic enrichment can enhance fish community development on degraded coral reef habitat. *Nature Communications*, 10(1), 1–7.

Harari, Y. N. (2014). *Sapiens: A brief history of humankind*. Random House.

Hughes, T. P., Barnes, M. L., Bellwood, D. R., Cinner, J. E., Cumming, G. S., Jackson, J. B. et al. (2017). Coral reefs in the Anthropocene. *Nature*, 546(7656), 82–90.

Kaplan, M. B., Mooney, T. A., Partan, J., & Solow, A. R. (2015). Coral reef species assemblages are associated with ambient soundscapes. *Marine Ecology Progress Series*, 533, 93–107.

McCauley, D. J., Gellner, G., Martinez, N. D., Williams, R. J., Sandin, S. A., Micheli, F. et al. (2018). On the prevalence and dynamics of inverted trophic pyramids and otherwise top-heavy communities. *Ecology Letters*, 21(3), 439–454.

Pandolfi, J. M., Bradbury, R. H., Sala, E., Hughes, T. P., Bjorndal, K. A., Cooke, R. G. et al. (2003). Global trajectories of the long-term decline of coral reef ecosystems. *Science*, 301(5635), 955–958.

Pandolfi, J. M., Connolly, S. R., Marshall, D. J., & Cohen, A. L. (2011). Projecting coral reef futures under global warming and ocean acidification. *Science*, 333(6041), 418–422.

Saltré, F., Rodríguez-Rey, M., Brook, B. W., Johnson, C. N., Turney, C. S., Alroy, J. et al. (2016). Climate change not to blame for late Quaternary megafauna extinctions in Australia. *Nature Communications*, 7(1), 1–7.

Singh, A., Wang, H., Morrison, W., & Weiss, H. (2008). *Fish biomass structure at pristine coral reefs and degradation by fishing*. arXiv preprint.

Singh, A., Wang, H., Morrison, W., & Weiss, H. (2012). Modeling fish biomass structure at near pristine coral reefs and degradation by fishing. *Journal of Biological Systems*, 20(01), 21–36.

Vermeij, M. J., Marhaver, K. L., Huijbers, C. M., Nagelkerken, I., & Simpson, S. D. (2010). Coral larvae move toward reef sounds. *PLoS One*, 5(5), e10660.

Westaway, M. C., Olley, J., & Grün, R. (2017). At least 17,000 years of coexistence: Modern humans and megafauna at the Willandra Lakes, South-Eastern Australia. *Quaternary Science Reviews*, 157, 206–211.

Coral Colonies: Do They Constitute Superorganisms?

Bayer, T., Neave, M. J., Alsheikh-Hussain, A., Aranda, M., Yum, L. K., Mincer, T. et al. (2013). The microbiome of the Red Sea coral *Stylophora pistillata* is dominated by tissue-associated Endozoicomonas bacteria. *Applied and Environmental Microbiology*, 79(15), 4759–4762.

Bordenstein, S. R., & Theis, K. R. (2015). Host biology in light of the microbiome: Ten principles of holobionts and hologenomes. *PLoS Biol*, 13(8), e1002226.

Carter, A. (2013). Coral's indispensable bacterial buddies. *Oceanus*, 50(2), 6.

Duffy, J. E. (1996). Eusociality in a coral-reef shrimp. *Nature*, 381(6582), 512–514.

Gateno, D., Israel, A., Barki, Y., & Rinkevich, B. (1998). Gastrovascular circulation in an octocoral: Evidence of significant transport of coral and symbiont cells. *The Biological Bulletin*, 194(2), 178–186.

Gordon, J., Knowlton, N., Relman, D. A., Rohwer, F., & Youle, M. (2013). Superorganisms and holobionts. *Microbe*, 8(4), 152–153.

Gould, S. J. (1985). *The flamingo's smile: Reflections in natural history*. WW Norton & Company.

Hatcher, B. G. (1997). Coral reef ecosystems: How much greater is the whole than the sum of the parts? *Coral Reefs*, 16(1), S77–S91.

Krakauer, D., Bertschinger, N., Olbrich, E., Flack, J. C., & Ay, N. (2020). The information theory of individuality. *Theory in Biosciences*, 139(2), 209–223.

Lovelock, J. (1986). Gaia: the world as living organism. *New Scientist* (1971), 112(1539), 25–28.

Mann, C. (1991). Lynn Margulis: Science's unruly earth mother. *Science*, 252(5004), 378–382.

Margulis, L., & Sagan, D. (1997). *Microcosmos: Four billion years of microbial evolution*. University of California Press.

Munro, C., Vue, Z., Behringer, R. R., & Dunn, C. W. (2019). Morphology and development of the Portuguese man of war, *Physalia physalis*. *Scientific Reports*, 9(1), 1–12.

Oren, U., Benayahu, Y., Lubinevsky, H., & Loya, Y. (2001). Colony integration during regeneration in the stony coral *Favia favus*. *Ecology*, 82(3), 802–813.

Reece, J. B., Urry, L. A., Cain, M. L., Wasserman, S. A., Minorsky, P. V., & Jackson, R. B. (2014). *Campbell Biology*. Pearson.

Ruxton, G. D., Humphries, S., Morrell, L. J., & Wilkinson, D. M. (2014). Why is eusociality an almost exclusively terrestrial phenomenon? *Journal of Animal Ecology*, 83(6), 1248–1255.
Shavit, A. (2004). Shifting values partly explain the debate over group selection. *Studies in History and Philosophy of Science Part C: Studies in History and Philosophy of Biological and Biomedical Sciences*, 35(4), 697–720.
Turak, E., & DeVantier, L. (2003). Reef-building corals of Bunaken National Park, North Sulawesi, Indonesia: Rapid ecological assessment of biodiversity and status. *Final Report to the International Ocean Institute Regional Centre for Australia & the Western Pacific.*
Williams, G. C., Hoeksema, B. W., & van Ofwegen, L. P. (2012). A fifth morphological polyp in pennatulacean octocorals, with a review of polyp polymorphism in the genera Pennatula and Pteroeides (Anthozoa: Pennatulidae). *Zoological Studies*, 51(7), 1006–1017.
Wilson, D. S., & Sober, E. (1989). Reviving the superorganism. *Journal of theoretical Biology*, 136(3), 337–356.

Resolving Darwin's Paradox

Brandl, S. J., Tornabene, L., Goatley, C. H., Casey, J. M., Morais, R. A., Côté, I. M. et al. (2019). Demographic dynamics of the smallest marine vertebrates fuel coral reef ecosystem functioning. *Science*, 364(6446), 1189–1192.
Cox, B. (2013). *Wonders of Life: Exploring the Most Extraordinary Phenomenon in the Universe*. Harper Collins.
De Goeij, J. M., Van Oevelen, D., Vermeij, M. J., Osinga, R., Middelburg, J. J., De Goeij, A. F., & Admiraal, W. (2013). Surviving in a marine desert: The sponge loop retains resources within coral reefs. *Science*, 342(6154), 108–110.
Gove, J. M., McManus, M. A., Neuheimer, A. B., Polovina, J. J., Drazen, J. C., Smith, C. R. et al. (2016). Near-island biological hotspots in barren ocean basins. *Nature Communications*, 7, 10581.
Primack, R. B. (1993). *Essentials of conservation biology* (Vol. 23). Sinauer Associates.
Schrodinger, E. (2012). *What is life?: With mind and matter and autobiographical sketches*. Cambridge University Press. (Original work published 1944.)

Chapter 3
The Reef Kaleidoscopic View – Why Are Coral Reefs So Colorful?

Huxley, A. (1956). *Heaven and hell*. Chatto & Windus.

The Constraints of Natural Selection

Alonso, W. J. (2015). Evolution of bright colours in animals: Worlds of prohibition and oblivion. *F1000Research*, 4.
Castaneda, C. (1968). *The teachings of Don Juan: A Yaqui way of knowledge*. University of California Press.
Cortesi, F., & Cheney, K. L. (2010). Conspicuousness is correlated with toxicity in marine opisthobranchs. *Journal of Evolutionary Biology*, 23(7), 1509–1518.
Marshall, N. J., Cortesi, F., de Busserolles, F., Siebeck, U. E., & Cheney, K. L. (2019). Colours and colour vision in reef fishes: Past, present and future research directions. *Journal of Fish Biology*, 95(1), 5–38.
Valdes, A., Ornelas-Gatdula, E., & Dupont, A. (2013). Color pattern variation in a shallow-water species of opisthobranch mollusc. *The Biological Bulletin*, 224(1), 35–46.
Wallace, A. R. (1879). The protective colours of animals. *Science*, 2, 128–137.
Watson, S. A., Morley, S. A., & Peck, L. S. (2017). Latitudinal trends in shell production cost from the tropics to the poles. *Science Advances*, 3(9), e1701362.

The Properties of Light Underwater

Helfman, G., Collette, B. B., Facey, D. E., & Bowen, B. W. (2009). *The diversity of fishes: biology, evolution, and ecology*. John Wiley & Sons.
Losey, G. S., Cronin, T. W., Goldsmith, T. H., Hyde, D., Marshall, N. J., & McFarland, W. N. (1999). The UV visual world of fishes: A review. *Journal of Fish Biology*, 54(5), 921–943.
Marshall, N. J., & Cheney, K. (2011). Color vision and color communication in reef fish. In A. P. Farrell (ed.), *Encyclopedia of fish physiology: from genome to environment*. Academic Press.
Marshall, N. J., Cortesi, F., de Busserolles, F., Siebeck, U. E., & Cheney, K. L. (2019). Colours and colour vision in reef fishes: Past, present and future research directions. *Journal of Fish Biology*, 95(1), 5–38.

The Principle of Camouflage

Behrens, R. R. (2008). Revisiting Abbott Thayer: non-scientific reflections about camouflage in art, war and zoology. *Philosophical Transactions of the Royal Society B: Biological Sciences*, 364(1516), 497–501.
Marshall, J. (1998). Why are reef fish so colorful? Fish residing around corals are living rainbows. But the biological utility of those hues is complex. *Scientific American*, (3), 54–57.
Pietsch, T. W., & Arnold, R. J. (2020). *Frogfishes: Biodiversity, Zoogeography, and Behavioral Ecology*. JHU Press.
Pietsch, T. W., & Grobecker, D. B. (1990). Frogfishes. *Scientific American*, 262(6), 96–103.
Randall, J. E. (2005). A review of mimicry in marine fishes. *Zoological Studies Tapei*, 44(3), 299.
Rowland, H. M. (2011). The history, theory and evidence for a cryptic function of countershading. In M. Stevens & S. Merilaita (eds.), *Animal camouflage: mechanisms and function*. Cambridge University Press, 53–72.
Schwab, I. R., & Marshall, J. (2003). Hiding in plain view. *British Journal of Ophthalmology*, 87(3), 262.
Shepherd, B., Wandell, M., & Ross, R. (2017). Mating, birth, larval development and settlement of Bargibant's pygmy seahorse, *Hippocampus bargibanti* (Syngnathidae), in aquaria. *Aquaculture, Aquarium, Conservation & Legislation*, 10(5), 1049–1063.
Smith, R. E., Grutter, A. S., & Tibbetts, I. R. (2012). Extreme habitat specialisation and population structure of two gorgonian-associated pygmy seahorses. *Marine Ecology Progress Series*, 444, 195–206.
Stevens, M., & Merilaita, S. (eds.). (2011). *Animal camouflage: mechanisms and function*. Cambridge University Press.
Thayer, A. H. (1896). The law which underlies protective coloration. *The Auk*, 13(2), 124–129.
Zuberbühler, T., 2018. *Frogfishes: Southeast Asia, Maldives, Red Sea*, http://www.critter.ch/frogfish-book.html rRetrieved March 5, 2021).

Human-Biased Point of View

Bompas, A., Kendall, G., & Sumner, P. (2013). Spotting fruit versus picking fruit as the selective advantage of human colour vision. *i-Perception*, 4(2), 84–94.
Darwin, C. R. 1872. *The expression of the emotions in man and animals* (1st edn). John Murray.
Marshall, N. J., Cortesi, F., de Busserolles, F., Siebeck, U. E., & Cheney, K. L. (2019). Colours and colour vision in reef fishes: Past, present and future research directions. *Journal of Fish Biology*, 95(1), 5–38.

Chapter 4

Illumination, Disguise and Vision Mechanisms in Coral Reefs

Bessho-Uehara, M., Yamamoto, N., Shigenobu, S., Mori, H., Kuwata, K., & Oba, Y. (2020). Kleptoprotein bioluminescence: Parapriacanthus fish obtain luciferase from ostracod prey. *Science Advances*, 6(2), eaax4942.

Biological Luminescence Mechanisms

Blake, W. (2013). *Delphi Complete Works of William Blake* (illustrated). Delphi Classics. (Original work published 1788.)

Clarke, A. C. (2011). *Dolphin Island.* Hachette UK. (Original work published 1963.)

De Brauwer, M., & Hobbs, J. P. A. (2016). Stars and stripes: Biofluorescent lures in the striated frogfish indicate role in aggressive mimicry. *Coral Reefs*, 35(4), 1171.

Dougherty, L. (2016). Flashing in the "Disco" Clam *Ctenoides ales* (Finlay, 1927): Mechanisms and Behavioral Function (Doctoral dissertation, UC Berkeley).

Harvey, E. N. (1957). *A history of luminescence from the earliest times until 1900.* American Philosophical Society.

Goethe, Johann, W. (1970). *Theory of colours.* MIT Press, 40, 46. (Original work published 1810.)

Mäthger, L. M., Bell, G. R., Kuzirian, A. M., Allen, J. J., & Hanlon, R. T. (2012). How does the blue-ringed octopus (*Hapalochlaena lunulata*) flash its blue rings? *Journal of Experimental Biology*, 215(21), 3752–3757.

Mazel, C. H., Cronin, T. W., Caldwell, R. L., & Marshall, N. J. (2004). Fluorescent enhancement of signaling in a mantis shrimp. *Science*, 303(5654), 51.

Naguit, M. A. A., Plata, K. C., Abisado, R. G., & Calugay, R. J. (2014). Evidence of bacterial bioluminescence in a Philippine squid and octopus hosts. *Aquaculture, Aquarium, Conservation & Legislation*, 7(6), 497–507.

Park, H. B., Lam, Y. C., Gaffney, J. P., Weaver, J. C., Krivoshik, S. R., Hamchand, R.et al. (2019). Bright green biofluorescence in sharks derives from bromo-kynurenine metabolism. *Iscience*, 19, 1291–1336.

Sparks, J. S., Schelly, R. C., Smith, W. L., Davis, M. P., Tchernov, D., Pieribone, V. A., & Gruber, D. F. (2014). The covert world of fish biofluorescence: A phylogenetically widespread and phenotypically variable phenomenon. *PLoS One*, 9(1), e83259.

Reef Dwellers' Vision System and Methods of Illumination

Cox, B. (2013). *Wonders of Life: Exploring the Most Extraordinary Phenomenon in the Universe.* Harper Collins.

Knight, K. (2010). Sea urchins use whole bogy as eye. *Journal of Experimental Biology*, 213(2), i–ii.

Shashar, N., Johnsen, S., Lerner, A., Sabbah, S., Chiao, C. C., Mäthger, L. M., & Hanlon, R. T. (2011). Underwater linear polarization: Physical limitations to biological functions. *Philosophical Transactions of the Royal Society B: Biological Sciences*, 366(1565), 649–654.

Yerramilli, D., & Johnsen, S. (2010). Spatial vision in the purple sea urchin *Strongylocentrotus purpuratus* (Echinoidea). *Journal of Experimental Biology*, 213(2), 249–255.

The Incredible Mystery of the Mantis Shrimp's Eye

Cox, B. (2013). *Wonders of Life: Exploring the Most Extraordinary Phenomenon in the Universe.* Harper Collins.

Cronin, T. W., Chiou, T. H., Caldwell, R. L., Roberts, N., & Marshall, J. (2009, August). Polarization signals in mantis shrimps. In J. A. Shaw (ed.), *Polarization Science and Remote Sensing IV* (Vol. 7461). International Society for Optics and Photonics, 74610C.

Patel, R. N., & Cronin, T. W. (2020). Mantis shrimp navigate home using celestial and idiothetic path integration. *Current Biology*, 30. 1981–1987.

Thoen, H. H., How, M. J., Chiou, T. H., & Marshall, J. (2014). A different form of color vision in mantis shrimp. *Science*, 343(6169), 411–413.

Octopuses: The Colorblind Masters of Disguise

Hanlon, R. (2007). Cephalopod dynamic camouflage. *Current Biology*, 17(11), R400–R404.

Ramirez, M. D., & Oakley, T. H. (2015). Eye-independent, light-activated chromatophore expansion (LACE) and expression of phototransduction genes in the skin of Octopus bimaculoides. *Journal of Experimental Biology*, 218(10), 1513–1520.

Stubbs, A. L., & Stubbs, C. W. (2016). Spectral discrimination in color blind animals via chromatic aberration and pupil shape. *Proceedings of the National Academy of Sciences*, 113(29), 8206–8211.

Verne, J. (1998). *Twenty thousand leagues under the sea*. Oxford University Press. (Original work published 1869.)

Chapter 5
Mutual Aid – Coral Reefs as a Symbiotic Society

Cahm, C. (2002). *Kropotkin: And the Rise of Revolutionary Anarchism, 1872–1886*. Cambridge University Press.

Gould, S. J. (1988). Kropotkin was no crackpot. *Natural History*, 7(97), 12–21.

Harman, O. (2011). *The price of altruism: George Price and the search for the origins of kindness*. WW Norton & Company.

Kropotkin, P. (1955). *Mutual Aid: A Factor of Evolution*. Anarchy Archives. (Original work published 1902.)

Kropotkin, P., & Walter, N. (2010). *Memoirs of a Revolutionist*. Courier Corporation. (Original work published 1899).

Margulis, L., & Sagan, D. (1997). *Microcosmos: Four billion years of microbial evolution*. University of California Press.

Natan, E., Fitak, R. R., Werber, Y., & Vortman, Y. (2020). *Symbiotic magnetic sensing: raising evidence and beyond*. *Philosophical Transactions of the Royal Society B*, 375 20190595.

What is Symbiosis?

Leung, T. L. F., & Poulin, R. (2008). Parasitism, commensalism, and mutualism: Exploring the many shades of symbioses. *Vie et Milieu*, 58(2), 107.

The Centrality of Coral's Symbiosis with Algae

Dubinsky, Z., & Stambler, N. (eds.). (2010). *Coral reefs: an ecosystem in transition*. Springer Science & Business Media.

Cleaning Stations: How Did They Evolve?

Axelrod, R., & Hamilton, W. D. (1981). The evolution of cooperation. *Science*, 211(4489), 1390–1396.

Bshary, R., & Würth, M. (2001). Cleaner fish *Labroides dimidiatus* manipulate client reef fish by providing tactile stimulation. *Proceedings of the Royal Society of London. Series B: Biological Sciences*, 268(1475), 1495–1501.

Grutter, A. (1996). Parasite removal rates by the cleaner wrasse *Labroides dimidiatus*. *Marine Ecology Progress Series*, 130, 61–70.

Grutter, A. S. (2010). Cleaner fish. *CurrentBbiology*, 20(13).

Poulin, R., & Grutter, A. S. (1996). Cleaning symbioses: Proximate and adaptive explanations. *Bioscience*, 46(7), 512–517.

Raihani, N. J., Grutter, A. S., & Bshary, R. (2010). Punishers benefit from third-party punishment in fish. *Science*, 327(5962), 171.

Soares, M. C., Oliveira, R. F., Ros, A. F., Grutter, A. S., & Bshary, R. (2011). Tactile stimulation lowers stress in fish. *Nature Communications*, 2(1), 1–5.

Trivers, R. L. (1971). The evolution of reciprocal altruism. *The Quarterly Review of Biology*, 46(1), 35–57.

The Sea Anemone and Its Partners

Elliott, J. K., & Mariscal, R. N. (1997). Acclimation or innate protection of anemonefishes from sea anemones? *Copeia*, 284–289.

Fautin, D. G. (1991). The anemonefish symbiosis: What is known and what is not. *Symbiosis*, 10, 23–46.

Randall, J. E., & Diamant, A. (2017). Examples of symbiosis in tropical marine fishes. *Journal of the Ocean Science Foundation*, 26, 95–115.

Examples of Symbiosis with Crustaceans in the Reef

Horká, I., De Grave, S., Fransen, C. H., Petrusek, A., & Ďuriš, Z. (2016). Multiple host switching events shape the evolution of symbiotic palaemonid shrimps (Crustacea: Decapoda). *Scientific Reports*, 6, 26486.

Kohda, M., Yamanouchi, H., Hirata, T., Satoh, S., & Ota, K. (2017). A novel aspect of goby–shrimp symbiosis: Gobies provide droppings in their burrows as vital food for their partner shrimps. *Marine Biology*, 164(1), 22.

Schnytzer, Y., Giman, Y., Karplus, I., & Achituv, Y. (2017). Boxer crabs induce asexual reproduction of their associated sea anemones by splitting and intraspecific theft. *PeerJ*, 5, e2954.

Chapter 6
Nature, Red in Tooth and Claw – Defense and Preying Mechanisms

Bengtson, S. (2002). Origins and early evolution of predation. *The Paleontological Society Papers*, 8, 289–318.

Dawkins, R. (1996). *The blind watchmaker: Why the evidence of evolution reveals a universe without design*. WW Norton & Company.

de Nooijer, S., Holland, B. R., & Penny, D. (2009). The emergence of predators in early life: There was no Garden of Eden. *PloS One*, 4(6), e5507.

Diogenes, L. (1959). *Lives of eminent philosophers*. Harvard University Press. (Original work compiled 3rd third century BC.)

Meyers, M. A., Chen, P. Y., Lin, A. Y. M., & Seki, Y. (2008). Biological materials: Structure and mechanical properties. *Progress in Materials Science*, 53(1), 1–206.

Tennyson, A. L. (1850). *In Memoriam A.H.H.* E. Moxon.

Cnidaria Preying Mechanism

Berking, S., & Herrmann, K. (2006). Formation and discharge of nematocysts is controlled by a proton gradient across the cyst membrane. *Helgoland Marine Research*, 60(3), 180–188.

Cannon, Q., & Wagner, E. (2003). Comparison of discharge mechanisms of Cnidarian cnidae and Myxozoan polar capsules. *Reviews in Fisheries Science*, 11(3), 185–219.

Lotan, A., Fishman, L., Loya, Y., & Zlotkin, E. (1995). Delivery of a nematocyst toxin. *Nature*, 375(6531), 456.

Nüchter, T., Benoit, M., Engel, U., Özbek, S., & Holstein, T. W. (2006). Nanosecond-scale kinetics of nematocyst discharge. *Current Biology*, 16(9), R316–R318.

Özbek, S., Balasubramanian, P. G., & Holstein, T. W. (2009). Cnidocyst structure and the biomechanics of discharge. *Toxicon*, 54(8), 1038–1045.

Plachetzki, D. C., Fong, C. R., & Oakley, T. H. (2012). Cnidocyte discharge is regulated by light and opsin-mediated phototransduction. *BMC Biology*, 10(1), 1–10.

The Deadly Strike of the Mantis Shrimp

Patek, S. N., & Caldwell, R. L. (2005). Extreme impact and cavitation forces of a biological hammer: Strike forces of the peacock mantis shrimp *Odontodactylus scyllarus*. *Journal of Experimental Biology*, 208(19), 3655–3664.

Patek, S. N., Korff, W. L., & Caldwell, R. L. (2004). Deadly strike mechanism of a mantis shrimp. *Nature*, 428(6985), 819–820.

Tadayon, M., Amini, S., Wang, Z., & Miserez, A. (2018). Biomechanical design of the mantis shrimp saddle: A biomineralized spring used for rapid raptorial strikes. *iScience*, 8, 271–282.

Van Der Wal, C., Ahyong, S. T., Ho, S. Y., & Lo, N. (2017). The evolutionary history of Stomatopoda (Crustacea: Malacostraca) inferred from molecular data. *PeerJ*, 5, e3844.

The Occult Power of Electric Rays (Torpedo Fishes)

Alves-Gomes, J. A. (2001). The evolution of electroreception and bioelectrogenesis in teleost fish: A phylogenetic perspective. *Journal of Fish Biology*, 58(6), 1489–1511.

Copenhaver, B. P. (1991). A tale of two fishes: Magical objects in natural history from antiquity through the scientific revolution. *Journal of the History of Ideas*, 373–398.

Piccolino, M. (2007). The taming of the electric ray: from a wonderful and dreadful "art" to "animal electricity" and "electric battery". In H. Whitaker, C. U. M. Smith & S. Finger (eds.), *Brain, Mind and Medicine: Essays in Eighteenth-Century Neuroscience*. Springer, 125–143.

Shapeshifters: The Dynamic Mimicry of the Mimic Octopus

Baker, B. (2010). Unusual adaptations: Evolution of the mimic octopus. *Bioscience*, 60(11), 962.

Castaneda, C. (1968). *The teachings of Don Juan: A Yaqui way of knowledge*. University of California Press.

Huffard, C. L., Saarman, N., Hamilton, H., & Simison, W. B. (2010). The evolution of conspicuous facultative mimicry in octopuses: An example of secondary adaptation? *Biological Journal of the Linnean Society*, 101(1), 68–77.

Norman, M. D., Finn, J., & Tregenza, T. (2001). Dynamic mimicry in an Indo–Malayan octopus. *Proceedings of the Royal Society of London. Series B: Biological Sciences*, 268(1478), 1755–1758.

The Aggressive Mimicry and Fast Gulp of Frogfishes

De Brauwer, M., & Hobbs, J. P. A. (2016). Stars and stripes: biofluorescent lures in the striated frogfish indicate role in aggressive mimicry. *Coral Reefs*, 35(4), 1171.

Pietsch, T. W., & Arnold, R. J. (2020). *Frogfishes: Biodiversity, Zoogeography, and Behavioral Ecology*. JHU Press.

Pietsch, T. W., & Grobecker, D. B. (1990). Frogfishes. *Scientific American*, 262(6), 96–103.

Zuberbühler T., 2018. *Frogfishes: Southeast Asia, Maldives, Red Sea*, http://www.critter.ch/frogfish-book.html (retrieved March 5, 2021).

Venomous Coral Reef Creatures

Aguado, F., & Marin, A. (2007). Warning coloration associated with nematocyst-based defences in aeolidiodean nudibranchs. *Journal of Molluscan Studies*, 73(1), 23–28.

Cheney, K. L., White, A., Mudianta, I. W., Winters, A. E., Quezada, M., Capon, R. J. et al. (2016). Choose your weaponry: Selective storage of a single toxic compound, latrunculin A, by closely related nudibranch molluscs. *PLoS One*, 11(1), e0145134.

Harris, R. J., & Jenner, R. A. (2019). Evolutionary ecology of fish venom: Adaptations and consequences of evolving a venom system. *Toxins*, 11(2), 60.

Martin, R., & Walther, P. (2002). Effects of discharging nematocysts when an eolid nudibranch feeds on a hydroid. *Journal of the Marine Biological Association of the United Kingdom*, 82(3), 455–462.

Stonik, V. A., & Stonik, I. V. (2014). Toxins produced by marine invertebrate and vertebrate animals: A short review. In P. Gopalakrishnakone et al. (eds.), *Marine and Freshwater Toxins*. Springer, 405–419.

Winters, A. E., Wilson, N. G., van den Berg, C. P., How, M. J., Endler, J. A., Marshall, N. J. et al. (2018). Toxicity and taste: Unequal chemical defences in a mimicry ring. *Proceedings of the Royal Society B: Biological Sciences*, 285(1880), 20180457.

Ziegman, R., & Alewood, P. (2015). Bioactive components in fish venoms. *Toxins*, 7(5), 1497–1531.

Zombie Powder: TTX Venom Mystery

Booth, W. (1988). Voodoo science. *Science*, 240(4850), 274.

Cook, J. (1777). A *voyage towards the South Pole and around the world performed in His Majesty's ships the "Resolution" and "Adventure" in the years 1772–75*. Shanan and Cadell.

Davis, E. W. (1983). The ethnobiology of the Haitian zombie. *Journal of ethnopharmacology*, 9(1), 85–104.

Hahn, P. (2007). Dead man walking: Wade Davis and the secret of the zombie poison. *Biology online*. Available at: http://skeptoid. com/episodes/4262 (retrieved March 5, 2021).

Hanifin, C. T. (2010). The chemical and evolutionary ecology of tetrodotoxin (TTX) toxicity in terrestrial vertebrates. *Marine Drugs*, 8(3), 577–593.

Hopkins, J. (2019). *Extreme Cuisine: The Weird and Wonderful Foods That People Eat*. Tuttle Publishing.

Kao, C. Y., & Yasumoto, T. (1990). Tetrodotoxin in "zombie powder." *Toxicon*, 28(2), 129–132.

Littlewood, R., & Douyon, C. (1997). Clinical findings in three cases of zombification. *The Lancet*, 350(9084), 1094–1096.

Lorentz, M. N., Stokes, A. N., Rößler, D. C., & Lötters, S. (2016). Tetrodotoxin. *Current Biology*, 26(19), R870–R872.A

Magarlamov, T. Y., Melnikova, D. I., & Chernyshev, A. V. (2017). Tetrodotoxin-producing bacteria: Detection, distribution and migration of the toxin in aquatic systems. *Toxins*, 9(5), 166.

Crowd Behavior: Why Do Fish School?

Canetti, E. (1962). *Crowds and power*. Macmillan.
Castro, P., & Huber, M. (2007). *Marine Biology* (7th edn). McGraw-Hill Education.
Kasumyan, A. O., & Pavlov, D. S. (2018). Evolution of schooling behavior in fish. *Journal of Ichthyology*, 58(5), 670–678.
Larsson, M. (2012). Why do fish school? *Current Zoology*, 58(1), 116–128.
Schellinck, J., & White, T. (2011). A review of attraction and repulsion models of aggregation: Methods, findings and a discussion of model validation. *Ecological Modelling*, 222(11), 1897–1911.
Warren, W. H. (2018). Collective motion in human crowds. *Current Directions in Psychological Science*, 27(4), 232–240.

Chapter 7
Reproduction – The Wild Side of Sex in Coral Reefs

Graham, S. (2002). Sex, gender, and priest in South Sulawesi, Indonesia. *IIAS Newsletter*, 29, 27.
Pattanaik, D. (2014). *The man who was a woman and other queer tales from Hindu lore*. Routledge.
Peletz, M. G. (2009). *Gender pluralism: Southeast Asia since early modern times*. Routledge.
Roughgarden, J. (2013). *Evolution's rainbow: Diversity, gender, and sexuality in nature and people*. University of California Press.

Hermaphroditism: Sex Change in Fishes

Avise, J. C., & Mank, J. E. (2009). Evolutionary perspectives on hermaphroditism in fishes. *Sexual Development*, 3(2–3), 152–163.
Benvenuto, C., Coscia, I., Chopelet, J., Sala-Bozano, M., & Mariani, S. (2017). Ecological and evolutionary consequences of alternative sex-change pathways in fish. *Scientific Reports*, 7(1), 9084.
Clark, W. C. (1978). Hermaphroditism as a reproductive strategy for metazoans; some correlated benefits. *New Zealand Journal of Zoology*, 5(4), 769–780.
Kobayashi, Y., Nagahama, Y., & Nakamura, M. (2013). Diversity and plasticity of sex determination and differentiation in fishes. *Sexual Development*, 7(1–3), 115–125.
Munday, P. L., Buston, P. M., & Warner, R. R. (2006). Diversity and flexibility of sex-change strategies in animals. *Trends in Ecology & Evolution*, 21(2), 89–95.
Munday, P. L., Kuwamura, T., & Kroon, F. J. (2010). Bidirectional sex change in marine fishes. *Reproduction and Sexuality in Marine Fishes: Patterns and Processes*, 241–271.
Shapiro, D. Y. (1981). Behavioural changes of protogynous sex reversal in a coral reef fish in the laboratory. *Animal Behaviour*, 29(4), 1185–1198.
Todd, E. V., Liu, H., Muncaster, S., & Gemmell, N. J. (2016). Bending genders: The biology of natural sex change in fish. *Sexual Development*, 10(5–6), 223–241.
Todd, E. V., Ortega-Recalde, O., Liu, H., Lamm, M. S., Rutherford, K. M., Cross, H. et al. (2019). Stress, novel sex genes, and epigenetic reprogramming orchestrate socially controlled sex change. *Science Advances*, 5(7), eaaw7006.
Warner, R. R. (1984). Mating behavior and hermaphroditism in coral reef fishes. *American Scientist*, 72(2), 128–136.

Hybrids: Interspecies Breeding

Arnold, M. L. (2006). *Evolution through genetic exchange*. Oxford University Press.
Arnold, M. L. (2015). *Divergence with genetic exchange*. Oxford University Press.
Arnold, M. L., & Larson, E. J. (2004). Evolution's new look. The *Wilson Quarterly* (1976–), 28(4), 60–73.
Chan, W. Y., Peplow, L. M., & van Oppen, M. J. (2019). Interspecific gamete compatibility and hybrid larval fitness in reef-building corals: Implications for coral reef restoration. *Scientific Reports*, 9(1), 1–13.
Douglas, M. (1966). *Purity and Danger: An Analysis of the Concepts of Pollution and Taboo*. Ark.
De Queiroz, K. (2005). Ernst Mayr and the modern concept of species. *Proceedings of the National Academy of Sciences*, 102(suppl 1), 6600–6607.
Homer (2007) *The Iliad* (trans. I. Johnston). Richer Resources Publications.
Mallet, J. (2005). Hybridization as an invasion of the genome. *Trends in Ecology & Evolution*, 20(5), 229–237.
Mayr, E. (1999). *Systematics and the origin of species, from the viewpoint of a zoologist*. Harvard University Press. (Original work published 1942.)
Richards, Z. T., & Hobbs, J. P. A. (2015). Hybridisation on coral reefs and the conservation of evolutionary novelty. *Current Zoology*, 61(1), 132–145.
Roark, E. B., Guilderson, T. P., Dunbar, R. B., Fallon, S. J., & Mucciarone, D. A. (2009). Extreme longevity in proteinaceous deep-sea corals. *Proceedings of the National Academy of Sciences*, 106(13), 5204–5208.
Schechter, M. (2006). Liminal and hybrid aspects of intersemiotics. *Zeitschrift fur Kulturwissenschaften*, 16, 1–15.
Tea, Y. K., Hobbs, J. P. A., Vitelli, F., DiBattista, J. D., Ho, S. Y., & Lo, N. (2020). Angels in disguise: Sympatric hybridization in the marine angelfishes is widespread and occurs between deeply divergent lineages. *Proceedings of the Royal Society B*, 287(1932), 20201459.
Vollmer, S. V., & Palumbi, S. R. (2002). Hybridization and the evolution of reef coral diversity. *Science*, 296(5575), 2023–2025.
Wilson, E. O. (1999). *The diversity of life*. WW Norton & Company.

A Death Sentence: Octopus Sex

Amodio, P., Boeckle, M., Schnell, A. K., Ostojíc, L., Fiorito, G., & Clayton, N. S. (2019). Grow smart and die young: Why did cephalopods evolve intelligence? *Trends in Ecology & Evolution*, 34(1), 45–56.
Anderson, R. C., Wood, J. B., & Byrne, R. A. (2002). Octopus senescence: The beginning of the end. *Journal of Applied Animal Welfare Science*, 5(4), 275–283.
Homer (1997). *The Odyssey* (trans. R. Fagles). Penguin. (Original work published eighth century BCE.)
Maderspacher, F. (2016). Death and the octopus. *Current Biology*, 26(13), R543–R544.
Mather, J. A., Anderson, R. C., & Wood, J. B. (2013). *Octopus: the ocean's intelligent invertebrate*. Timber Press.
Morse, P., Huffard, C. L., Meekan, M. G., McCormick, M. I., & Zenger, K. R. (2018). Mating behaviour and postcopulatory fertilization patterns in the southern blue-ringed octopus, *Hapalochlaena maculosa*. *Animal Behaviour*, 136, 41–51.
Oakwood, M., Bradley, A. J., & Cockburn, A. (2001). Semelparity in a large marsupial. *Proceedings of the Royal Society of London. Series B: Biological Sciences*, 268(1465), 407–411.

Nudibranch Mating

Yonow, N. (2008). *Sea slugs of the Red Sea* (No. 74). Pensoft Pub.

Seahorse "feminism"

Lourie, S. A., Foster, S. J., Cooper, E. W., & Vincent, A. C. (2004). A guide to the identification of seahorses. *Project Seahorse and TRAFFIC North America*, 114.

Chapter 8
Evolutionary Themes

Dennett, D. C. (1996). *Darwin's Dangerous Idea: Evolution and the Meanings of Life*. Simon and Schuster.
Fujii, J. A., McLeish, D., Brooks, A. J., Gaskell, J., & Van Houtan, K. S. (2018). Limb-use by foraging marine turtles, an evolutionary perspective. *PeerJ*, 6, e4565.
Gary, R. (2019). *The Kites*. Penguin Classic.
Goff, P. (2017). Panpsychism is crazy, but it's also most probably true. *Aeon*, March, 1.
Gould, S. J., & Lewontin, R. C. (1979). The spandrels of San Marco and the Panglossian paradigm: A critique of the adaptationist programme. *Proceedings of the Royal Society of London. Series B. Biological Sciences*, 205(1161), 581–598.
Gould, S. J., & Vrba, E. S. (1982). Exaptation – a missing term in the science of form. *Paleobiology*, 8(1), 4–15.
Jablonka, E., & Lamb, M. J. (2014). *Evolution in four dimensions, revised edition: Genetic, epigenetic, behavioral, and symbolic variation in the history of life*. MIT Press.
Laland, K. N., Uller, T., Feldman, M. W., Sterelny, K., Müller, G. B., Moczek, A. et al. (2015). The extended evolutionary synthesis: Its structure, assumptions and predictions. *Proceedings of the Royal Society B: Biological Sciences*, 282(1813), 20151019.
Lyson, T. R., Rubidge, B. S., Scheyer, T. M., de Queiroz, K., Schachner, E. R., Smith, R. M. et al. (2016). Fossorial origin of the turtle shell. *Current Biology*, 26(14), 1887–1894.
Parmentier, E., Diogo, R., & Fine, M. L. (2017). Multiple exaptations leading to fish sound production. *Fish and Fisheries*, 18(5), 958–966.
Seilacher, A. (1972). Divaricate patterns in pelecypod shells. *Lethaia*, 5(3), 325–343.

Cephalopods and Animal Consciousness

Amodio, P., Boeckle, M., Schnell, A. K., Ostojíc, L., Fiorito, G., & Clayton, N. S. (2019). Grow smart and die young: Why did cephalopods evolve intelligence? *Trends in Ecology & Evolution*, 34(1), 45–56.
Birch, J., Burn, C., Schnell, A., Browning, H., & Crump, A. (2021). Review of the Evidence of Sentience in Cephalopod Molluscs and Decapod Crustaceans. Report published by LSE Consulting.
Bohm, D. (1990). A new theory of the relationship of mind and matter. *Philosophical Psychology*, 3(2–3), 271–286.
Brown, C. (2015). Fish intelligence, sentience and ethics. *Animal cognition*, 18(1), 1–17.
Cronin, T. W., Caldwell, R. L., & Marshall, J. (2006). Learning in stomatopod crustaceans. UMBC Faculty Collection.
Darwin, C. (1871). *The descent of man and selection in relation to sex* (Vol. 1). D. Appleton.
De Waal, F. (2016). *Are we smart enough to know how smart animals are?* WW Norton & Company.
Ginsburg, S., & Jablonka, E. (2019). *The evolution of the sensitive soul: Learning and the origins of consciousness*. MIT Press.
Godfrey-Smith, P. (2013). Cephalopods and the evolution of the mind. *Pacific Conservation Biology*, 19(1), 4–9.
Godfrey-Smith, P. (2016). *Other minds: The octopus, the sea, and the deep origins of consciousness*. Farrar, Straus and Giroux.

Gould, S. J. (2011). *Full house*. Harvard University Press.
Gross, C. (2010). Alfred Russell Wallace and the evolution of the human mind. *The Neuroscientist*, 16(5), 496–507.
Haralson, J. V., Groff, C. I., & Haralson, S. J. (1975). Classical conditioning in the sea anemone, *Cribrina xanthogrammica*. *Physiology & behavior*, 15(4), 455–460.
Horowitz, A. (2017). Smelling themselves: Dogs investigate their own odors longer when modified in an "olfactory mirror" test. *Behavioural processes*, 143, 17–24.
Kohda, M., Hotta, T., Takeyama, T., Awata, S., Tanaka, H., Asai, J. Y., & Jordan, A. L. (2019). If a fish can pass the mark test, what are the implications for consciousness and self-awareness testing in animals? *PLoS Biol*, 17(2).
Lorenz, K., Huxley, J., & Wilson, M. K. (1962). *King Solomon's ring: new light on animal ways*. Time Incorporated.
Nagel, T. (1974). What is it like to be a bat? *The Philosophical Review*, 83(4), 435–450.
Steinegger, M., Roche, D. G., & Bshary, R. (2018). Simple decision rules underlie collaborative hunting in yellow saddle goatfish. *Proceedings of the Royal Society B: Biological Sciences*, 285(1871), 20172488.
Sullivan, W. (1972). The Einstein papers: A man of many parts. *The New York Times*, March, 29.
Tononi, G., & Koch, C. (2015). Consciousness: here, there and everywhere? *Philosophical Transactions of the Royal Society B: Biological Sciences*, 370(1668), 20140167.
Wittgenstein, L. (2009). *Philosophical investigations*. John Wiley & Sons.

Origin of the Octopus: Terrestrial or Cosmic?

Albertin, C. B., Simakov, O., Mitros, T., Wang, Z. Y., Pungor, J. R., Edsinger-Gonzales, E. et al. (2015). The octopus genome and the evolution of cephalopod neural and morphological novelties. *Nature*, 524(7564), 220.
Douglas, K. M., Sutton, R. M., & Cichocka, A. (2017). The psychology of conspiracy theories. *Current Directions in Psychological Science*, 26(6), 538–542.
Fox, D. (2016). What sparked the Cambrian explosion? *Nature*, 530(7590), 268.
Ginsburg, S., & Jablonka, E. (2019). *The evolution of the sensitive soul: Learning and the origins of consciousness*. MIT Press.
Hoyle, F., & Wickramasinghe, C. (1979). *Diseases from space*. JM Dent & Sons.
Liscovitch-Brauer, N., Alon, S., Porath, H. T., Elstein, B., Unger, R., Ziv, T. et al. (2017). Trade-off between transcriptome plasticity and genome evolution in cephalopods. *Cell*, 169(2), 191–202.
Marshall, C. R. (2006). Explaining the Cambrian "explosion" of animals. *Annual Review of Earth and Planetary Sciences*, 34, 355–384.
Persson, D., Halberg, K. A., Jørgensen, A., Ricci, C., Møbjerg, N., & Kristensen, R. M. (2011). Extreme stress tolerance in tardigrades: Surviving space conditions in low earth orbit. *Journal of Zoological Systematics and Evolutionary Research*, 49, 90–97.
Sloan, D., Batista, R. A., & Loeb, A. (2017). The resilience of life to astrophysical events. *Scientific Reports*, 7(1), 5419.
Steele, E. J., Al-Mufti, S., Augustyn, K. A., Chandrajith, R., Coghlan, J. P., Coulson, S. G. et al. (2018). Cause of Cambrian Explosion – Terrestrial or cosmic? *Progress in biophysics and molecular biology*, 136, 3–23.
Von Daniken, E. (1968). *Chariots of the Gods*. Putnam.
Wickramasinghe, C. (2010). The astrobiological case for our cosmic ancestry. *International Journal of Astrobiology*, 9, 119–129.

The Evolution of Animal Eyes: Convergent or Divergent Evolution?

Birk, M. H., Blicher, M. E., & Garm, A. (2018). Deep-sea starfish from the Arctic have well-developed eyes in the dark. *Proceedings of the Royal Society B: Biological Sciences*, 285(1872), 20172743.

Cox, B. (2013). *Wonders of Life: Exploring the Most Extraordinary Phenomenon in the Universe.* Harper Collins.

Darwin, C. (1859). *On the Origin of Species by Means of Natural Selection Or the Preservation of Favoured Races in the Struggle for Life.* H. Milford; Oxford University Press.

Fernald, R. D. (2006). Casting a genetic light on the evolution of eyes. *Science*, 313(5795), 1914–1918.

Gehring, W. J. (2005). New perspectives on eye development and the evolution of eyes and photoreceptors. *Journal of Heredity*, 96(3), 171–184.

Land, M. F., & Nilsson, D. E. (2012). *Animal eyes.* Oxford University Press.

Palmer, B. A., Taylor, G. J., Brumfeld, V., Gur, D., Shemesh, M., Elad, N. et al. (2017). The image-forming mirror in the eye of the scallop. *Science*, 358(6367), 1172–1175.

Parker, A. (2016). *In the Blink of an Eye: How Vision Kick-started the Big Bang of Evolution.* Natural History Museum.

Yoshida, M. A., Yura, K., & Ogura, A. (2014). Cephalopod eye evolution was modulated by the acquisition of Pax-6 splicing variants. *Scientific Reports*, 4, 4256.

Better Red Than Dead: The Evolution of Warning Coloration

Briolat, E. S., Burdfield-Steel, E. R., Paul, S. C., Rönkä, K. H., Seymoure, B. M., Stankowich, T., & Stuckert, A. M. (2019). Diversity in warning coloration: Selective paradox or the norm? *Biological Reviews*, 94(2), 388–414.

Cortesi, F., & Cheney, K. L. (2010). Conspicuousness is correlated with toxicity in marine opisthobranchs. *Journal of Evolutionary Biology*, 23(7), 1509–1518.

Guilford, T., & Dawkins, M. S. (1993). Are warning colors handicaps? *Evolution*, 47(2), 400–416.

Hegna, R. H., Nokelainen, O., Hegna, J. R., & Mappes, J. (2013). To quiver or to shiver: Increased melanization benefits thermoregulation, but reduces warning signal efficacy in the wood tiger moth. *Proceedings of the Royal Society B: Biological Sciences*, 280(1755), 20122812.

Marples, N. M., Kelly, D. J., & Thomas, R. J. (2005). Perspective: The evolution of warning coloration is not paradoxical. *Evolution*, 59(5), 933–940.

Smith, J. M., & Harper, D. (2003). *Animal signals.* Oxford University Press.

Zahavi, A., & Zahavi, A. (1999). *The handicap principle: A missing piece of Darwin's puzzle.* Oxford University Press.

Chapter 9
Mathematical Beauty in Coral Reefs

Hawking, S. (2016). *Does God play dice?* (1999). https://www.hawking.org.uk/in-words/lectures/does-god-play-dice (retrieved March 6, 2021).

Vitruvius, P. (1901). *Vitruvius: The Ten Books on Architecture* (trans. M. Hicky). Forgotten Books.

Whitehead, A. N., Griffin, D. R. (Ed.), & Sherburne, D. W (Ed.). (1979). *Process and reality, an essay in cosmology.* Free Press. (Original work published 1927/28.)

The Divine Proportion

Corbusier, L. (1954). *The modulor: a harmonious measure to the human scale universally applicable to architecture and mechanics* (Vol. 1). Harvard University Press.

Livio, M. (2008). *The golden ratio: The story of phi, the world's most astonishing number.* Broadway Books.

Meyers, M. A., Chen, P. Y., Lin, A. Y. M., & Seki, Y. (2008). Biological materials: Structure and mechanical properties. *Progress in Materials Science*, 53(1), 1–206.

Olsen, S. (2006). *The golden section: nature's greatest secret.* Bloomsbury Publishing USA.

Salamone, M. A. (2019). The two supreme principles of Plato's Cosmos – the one and the indefinite dyad – the division of a straight line into extreme and mean ratio, and Pingala's Mātrāmeru. *Symmetry*, 11(1), 98.

Symmetry and the Evolution of Flatfishes

Finnerty, J. R. (2005). Did internal transport, rather than directed locomotion, favor the evolution of bilateral symmetry in animals? *BioEssays*, 27(11), 1174–1180.

Palestis, B. G., & Trivers, R. (2016). A longitudinal study of changes in fluctuating asymmetry with age in Jamaican youth. *Symmetry*, 8(11), 123.

Schreiber, A. M. (2013). Flatfish: An asymmetric perspective on metamorphosis. *Current Topics in Developmental Biology*, 103, 167–194).

The Hyperbolic Universe

Blake, W. (1972). *The pickering manuscript.* Pierpont Morgan Library. (Original work published 1803.)

Halsted, G. B. (1900). Non-Euclidean geometry. *The American Mathematical Monthly*, 7(5), 123–133.

Wallace, A. R. (1898). *The wonderful century: its successes and its failures.* G. N. Morang.

"Endless forms most beautiful" – Turing Patterns

Kondo, S., & Asai, R. (1995). A reaction–diffusion wave on the skin of the marine angelfish Pomacanthus. *Nature*, 376(6543), 765–768.

Meinhardt, H. (2009). *The algorithmic beauty of sea shells.* Springer Science & Business Media.

Metz, H. C., Manceau, M., & Hoekstra, H. E. (2011). Turing patterns: How the fish got its spots. *Pigment Cell & Melanoma Research*, 24(1), 12–14.

Stewart, I. (2011). *The mathematics of life.* Basic Books.

Turing, A. M. (1952). The chemical basis of morphogenesis. *Bulletin of mathematical biology*, 52(1–2), 153–197.

Coral Reef as Complex Systems

Bollati, E., D'Angelo, C., Alderdice, R., Pratchett, M., Ziegler, M., & Wiedenmann, J. (2020). Optical feedback loop involving dinoflagellate symbiont and scleractinian host drives colorful coral bleaching. *Current Biology*, 30, 2433–2445

Buddemeier, R. W., & Fautin, D. G. (1993). Coral bleaching as an adaptive mechanism. *Bioscience*, 43(5), 320–326.

Capra, F., & Luisi, P. L. (2014). *The systems view of life: A unifying vision.* Cambridge University Press.

Connell, J. H., Hughes, T. P., & Wallace, C. C. (1997). A 30-year study of coral abundance, recruitment, and disturbance at several scales in space and time. *Ecological Monographs*, 67(4), 461–488.

Cropp, R., Gabric, A., van Tran, D., Jones, G., Swan, H., & Butler, H. (2018). Coral reef aerosol emissions in response to irradiance stress in the Great Barrier Reef, Australia. *Ambio*, 47(6), 671–681.

Dizon, R. T., & Yap, H. T. (2006). Understanding coral reefs as complex systems: Degradation and prospects for recovery. *Scientia Marina*, 70(2), 219–226.

Gallagher, R., Appenzeller, T., & Normile, D. (1999). Beyond reductionism. *Science*, 284(5411), 79.

Gordon, J., Knowlton, N., Relman, D. A., Rohwer, F., & Youle, M. (2013). Superorganisms and holobionts. *Microbe*, 8(4), 152–153.
Jackson, R., Gabric, A., Cropp, R., & Woodhouse, M. (2019). Reviews and syntheses: Marine biogenic aerosols and the ecophysiology of coral reefs. *Biogeosciences Discussions*, 1–40.
Parrott, L. (2002). Complexity and the limits of ecological engineering. *Transactions of the ASAE*, 45(5), 1697.
Parrott, L., & Kok, R. (2000). Incorporating complexity in ecosystem modelling. *Complexity International*, 7, 1–19.
Rowan, R. (2004). Thermal adaptation in reef coral symbionts. *Nature*, 430(7001), 742.
Taylor, B. M., Benkwitt, C. E., Choat, H., Clements, K. D., Graham, N. A., & Meekan, M. G. (2020). Synchronous biological feedbacks in parrotfishes associated with pantropical coral bleaching. *Global Change Biology*, 26(3), 1285–1294.
Wiener, N. (2019). *Cybernetics or Control and Communication in the Animal and the Machine*. MIT Press. (Original work published 1961.)

Epilogue

Costanza, R., De Groot, R., Sutton, P., Van der Ploeg, S., Anderson, S. J., Kubiszewski, I. et al. (2014). Changes in the global value of ecosystem services. *Global Environmental Change*, 26, 152–158.
Denevan, W. M. (1992). The pristine myth: The landscape of the Americas in 1492. *Annals of the Association of American Geographers*, 82(3), 369–385.
Ginsburg, S., & Jablonka, E. (2019). *The evolution of the sensitive soul: Learning and the origins of consciousness*. MIT Press.
McCauley, D. J., Pinsky, M. L., Palumbi, S. R., Estes, J. A., Joyce, F. H., & Warner, R. R. (2015). Marine defaunation: Animal loss in the global ocean. *Science*, 347(6219), 1255641.
Naess, A. (1990). *Ecology, community and lifestyle: outline of an ecosophy*. Cambridge University Press.
Rieder, J. (2014). *Gospel of Freedom: Martin Luther King, Jr.'s Letter from Birmingham Jail and the Struggle that Changed a Nation*. Bloomsbury Publishing USA.
Stephens, L., Fuller, D., Boivin, N., Rick, T., Gauthier, N., Kay et al. (2019). Archaeological assessment reveals Earth's early transformation through land use. *Science*, 365(6456), 897–902.
Vulliamy E. (2009). Peter Singer. *The Guardian*, February 14.
White, L. (1967). The historical roots of our ecologic crisis. *Science*, 155(3767), 1203–1207.

Figure Credits

The following illustrations were based on vector images purchased from Shutterstock:

Typical anatomy of a stony coral – Designua/Shutterstock.com

Coral reproduction through broadcast spawning – Designua/Shutterstock.com

Global biodiversity of reef-building corals – KPPWC/Shutterstock.com

High-level anatomy of a Portuguese man o' war – sciencepics/Shutterstock.com

Chromatic aberration effect – Fouad A. Saad/Shutterstock.com

Typical cnidarian stinging cell – Aldona Griskeviciene/Shutterstock.com

Golden Spiral in a nautilus – CKA/Shutterstock.com

The following illustrations were modified and based on illustrations IPR free images from Freepik:

Typical cnidarian stinging cell – macrovector/Freepik

The following images are public domain and were retrieved from Wikipedia:

A Sunday Afternoon on the Island of La Grande Jatte (Seurat, G. [1884–1886]).

File: A Sunday on La Grande Jatte, Georges Seurat, 1884.jpg. (2020, August 1). Wikimedia Commons, the free media repository. Retrieved 11:41, April 2, 2021 from https://commons.wikimedia.org/w/index.php?title=File:A_Sunday_on_La_Grande_Jatte,_Georges_Seurat,_1884.jpg&oldid=436337042.

The School of Athens (Raffaello Sanzio da Urbino [1509–1511])

File: "The School of Athens" by Raffaello Sanzio da Urbino.jpg. (2021, March 21). Wikimedia Commons, the free media repository. Retrieved 11:46, April 2, 2021 from https://commons.wikimedia.org/w/index.php?title=File:%22The_School_of_Athens%22_by_Raffaello_Sanzio_da_Urbino.jpg&oldid=544927061.

>> A giant barrel sponge.
Puerto Galera, Philippines.

Index

List of Key Terms

List of People

< Golden trumpetfish against colorful reef background. Bunaken, Indonesia.

About the Author

Oded Degany has spent the last two decades exploring the secrets of tropical coral reefs. An M&A and corporate development executive by profession, Oded spends his free time satisfying his immense curiosity and exploring multiple disciplines that broadly address diverse life phenomena. Oded has a Bachelor of Physics from the Technion Israel Institute of Technology, an MBA from Tel Aviv University, and is a graduate (Cum Laude) of the Biological Thought program in Israel's Open University.

About the Book

Reflections Underwater is a unique, illuminating book that explores a dazzling variety of topics and concepts relating to coral reefs. The book speaks to nature-lovers who have an intellectual interest in the mesmerizing world of this unique habitat and aims to make the coral reef experience intellectually stimulating and engrossing. The book adopts a holistic, multidisciplinary perspective by weaving together scientific and humanistic ideas, including psychology, evolution, zoology, philosophy, mathematics, art, physics and more.

Reflections Underwater is an eclectic collection of essays that address topics and questions such as:

* Why are we so attracted to water and colorful reefs? Do coral reefs constitute a superorganism?
* What are the evolutionary drivers behind the extraordinarily high biodiversity in the Coral Triangle?
* Why are reefs so colorful?
* How did cleaning station symbiosis evolve?
* The non-conventional variety of reproduction strategies in reef animals
* Extraordinary prey and defense mechanisms in reef animals
* Why is sex a death sentence for octopuses?
* Do reef animals have consciousness?
* Is the origin of the octopus cosmic or terrestrial?
* How did warning coloration evolve?
* How does the golden ratio manifest itself in coral reefs?
* Understanding asymmetry in flatfish...

and much more.